RAIL GUN

John Batchelor/Ian Hogg

Charles Scribner's Sons, New York

A-4.73 (1)

Library of Congress Catalogue Card Number **72–12322** **SBN: 684–13342–3**

Produced by Davis and Harrison Visual Productions Limited,
London, WC2H 9PJ

Designed by Chris Harrison

Edited by S. L. Mayer

The Publishers wish to thank
Peter Sarson, Tony Bryan, Helen Downton and Gerhard Rumfeld
for their invaluable help in preparing the illustrations
for this book.

Acknowledgments

BUNDERS ARCHIV, COBLENZ, GERMANY.
LABORATOIRE CENTRALE DE L'ARMAMENT, FRANCE.
ROYAL ARMOURED CORPS MUSEUM, BOVINGDON, ENGLAND.
ROYAL ARTILLERY MUSEUM, WOOLWICH, ENGLAND.

Printed in Great Britain

CONTENTS

Rail Guns

The use of the railway train to bestow mobility on a piece of ordnance appears to have been slow to develop. If we take the introduction of a viable railway system as being 1825, when Stephenson's Stockton and Darlington Line was opened, almost thirty years went by before the appearance of the first recorded suggestion. However, examination of the other half of the idea—the artillery—reveals that these years were a time of absolutely no technical advance in that field either, and the inescapable conclusion is that there was little interest in military matters at the time, so that the possible connection escaped the usually alert eyes of the inventors of the day.

It was in 1853 that the first move was made. A Mr. Anderson published a pamphlet in Britain called 'National Defence' in which he 'proposed a complete plan of wrought-iron protected railway carriages'. Unfortunately it is not clear whether he actually advocated mounting artillery in these vehicles, or simply intended them as armored transport for reinforcing troops. Indeed, his ideas seem to have gone unnoticed by the authorities, since after reading of the use of armored trains in Egypt, his widow approached the War Office in 1882 drawing their attention to the pamphlet and requesting some tangible recognition of her husband's prior claim. But the War Office 'could not admit that anything done in Egypt was due to him'.

The first definite marriage between gun and railway car was during the American Civil War, when a smoothbore cast iron 13-inch mortar was mounted on a flatcar and taken into action in the campaign against Richmond. This mortar weighed some 17,000 pounds and the standard flatcar had to be heavily reinforced to withstand the firing blow. With the mortar secured to the car it was impossible to traverse the weapon in order to aim, and pointing was done by laying a short section of curved track and pushing the equipment along this until the barrel was laid in the desired direction. On firing, the mount was allowed a small recoil movement on the car, and the remainder of the recoil was absorbed by the whole affair trundling backwards until arrested by the crew leaping in to apply the brakes and push it back into place with handspikes.

This initial appearance of the railway gun is of particular interest, since it introduces not only the weapon but two features which became more important in later years and which are usually attributed to European inventors—firstly, the 'dual recoil' system—albeit in a primitive form—and secondly, the use of curved track to aim the weapon. This latter feature is so firmly believed to be of French origin that the length of curved track is almost always referred to by the French term 'Epi'.

This Federal innovation was shortly copied by the Confederate Army, notably by General Brent who mounted a number of various pieces on flatcars, shielded by railroad ties and sheet iron, and employed them on numerous occasions.

With the Civil War over, the railway artillery idea lapsed for awhile. Although the Franco-German border area was reasonably well-served by railway lines, and both participants in the War of 1870 used railways for troop movement and supply, there is no record of any attempt to put guns on the track. The next appearance of the rail gun was in Egypt in 1882 when, after the landings of the British Navy at Alexandria, it seems that Captain Scott, that indefatigable extemporiser of artillery, armored some wagons with boiler iron, placed some naval landing guns on open wagons, and used the guns in a protective role to accompany troop trains.

1
The Pioneer; the first recorded use of a rail-mounted gun was this 13-inch smoothbore mortar in the Richmond campaign during the American Civil War.

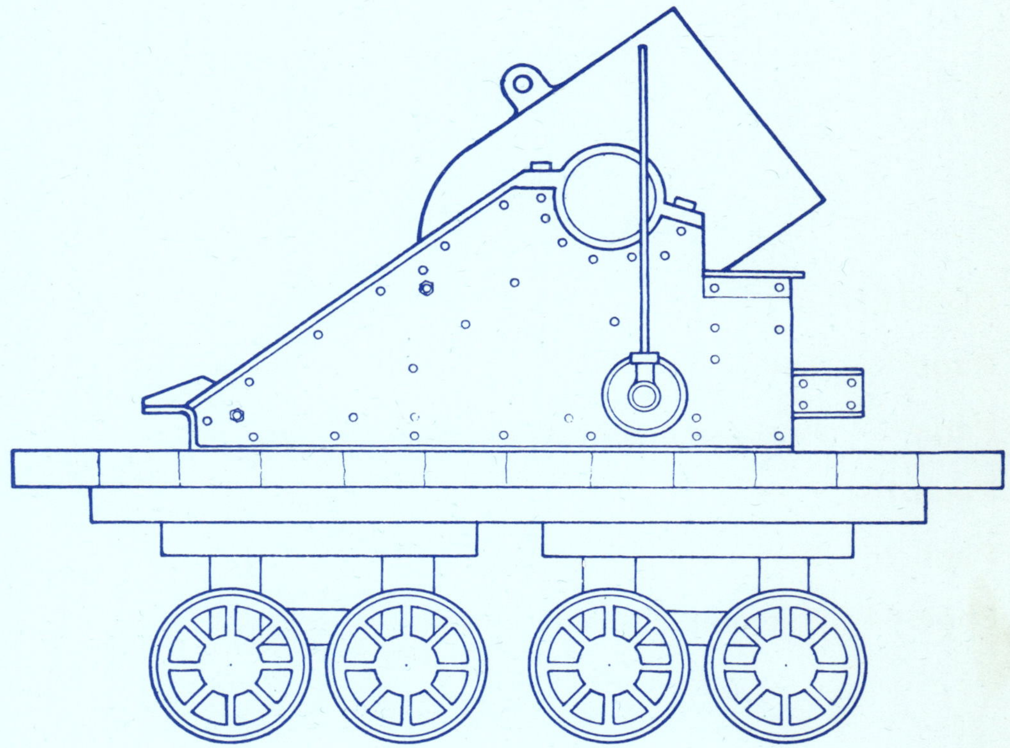

This application brings into focus the divergent purposes of placing guns on railway mountings and it will be as well to consider the point before going further. Railway guns, as generally considered, are large guns on special mountings which enable them to be moved from place to place by the normal rail network of a country, emplaced either on the normal track or on track specially laid for them, and fired, usually after some form of track preparation. It follows from this that the railway gun is essentially an offensive weapon, or at least a weapon which merely uses the railway line as a convenient medium for movement into its battle position.

On the other hand it is, as the Alexandria example shows, equally possible to place a gun of smaller caliber on a railway truck and use it to defend the train itself. This type of weapon usually forms part of an armored train, either for the movement of troops in a war zone or as a mobile patrol unit to keep a railway line open in an area liable to attack by irregular forces. Probably the most notable use of armored trains was by the Red Army during the Russian Revolution, when large areas had to be controlled and lines of communication kept open in the face of White

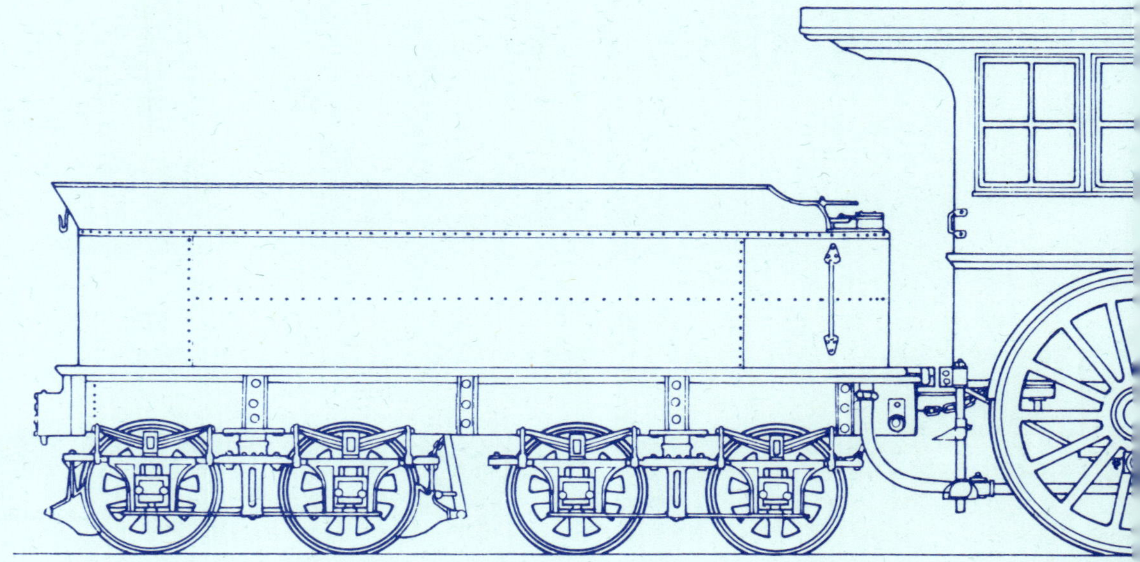

Russian infiltration. They were revived in that part of the world during the Second World War, when both the Germans and Russians found it necessary to provide protection against partisan and guerrilla activities. Britain also used this type of defensive train to carry light weapons and patrol the coastal railway lines as an anti-invasion measure in 1940–41, while the French had a number which were intended to supplement the Maginot Line defenses. But these devices, interesting as they are, are not railway artillery, and we cannot devote further space to them here.

Railway guns, by and large, were the weapon of the continental armies. Where a country has borders large distances apart, it becomes necessary for them to develop a highly mobile reserve of heavy guns which can be moved rapidly from one part of the country to another in order to face a threat or reinforce an attack. They also demand a well-developed rail network to enable them to move at will to cover any point, and a heavy engineering industry capable of designing and building them. In the circumstances it is not surprising to find the first move towards railway guns as they finally developed was in France. Yet the man who sowed the seed was an American.

General Brent, the ex-chief of Confederate Railway Artillery, had maintained his interest in the possibility of using railways for moving guns—one might almost say his faith—and in the late 1880s he managed to interest General Piegne of the French Army in his ideas. The events of 1870 were still fresh in French minds and there was a keenness to discover new methods and formulate new doctrines. Piegne appears to have had visions of the railway gun as a method of supplying fire with movement to support the future attacking policy of the French Army, and he began to advocate the construction of railway batteries. But the high command were not interested; the railway line was too confining. The spirit of 'audace' demanded the ability to move rapidly in any direction without the overriding consideration of having to go where the railway line went. And so the fast fire support of the new French Army was to be the light and maneuverable field gun, which eventually appeared as the 75mm Model of 1897.

But if the Army hadn't responded to Piegne's arguments, the gunmakers had. Piegne had taken his ideas to Schneider-Canet, and between them they produced a small number of 12cm guns on railway truck mountings. Even when confronted with a working weapon, the French Army were still not amused, and Schneider eventually sold these first weapons to the Danish Government for use as coastal defence weapons, an application which had attractions to countries with long coastlines and short purses.

General Piegne appears to have given up the struggle about this time, having failed to interest his own army, but Schneider had the share-

2
After the success of the 13-inch mortar, numerous other weapons, such as the muzzle-loader shown here, were flatcar-mounted and used by both Confederate and Union artillery

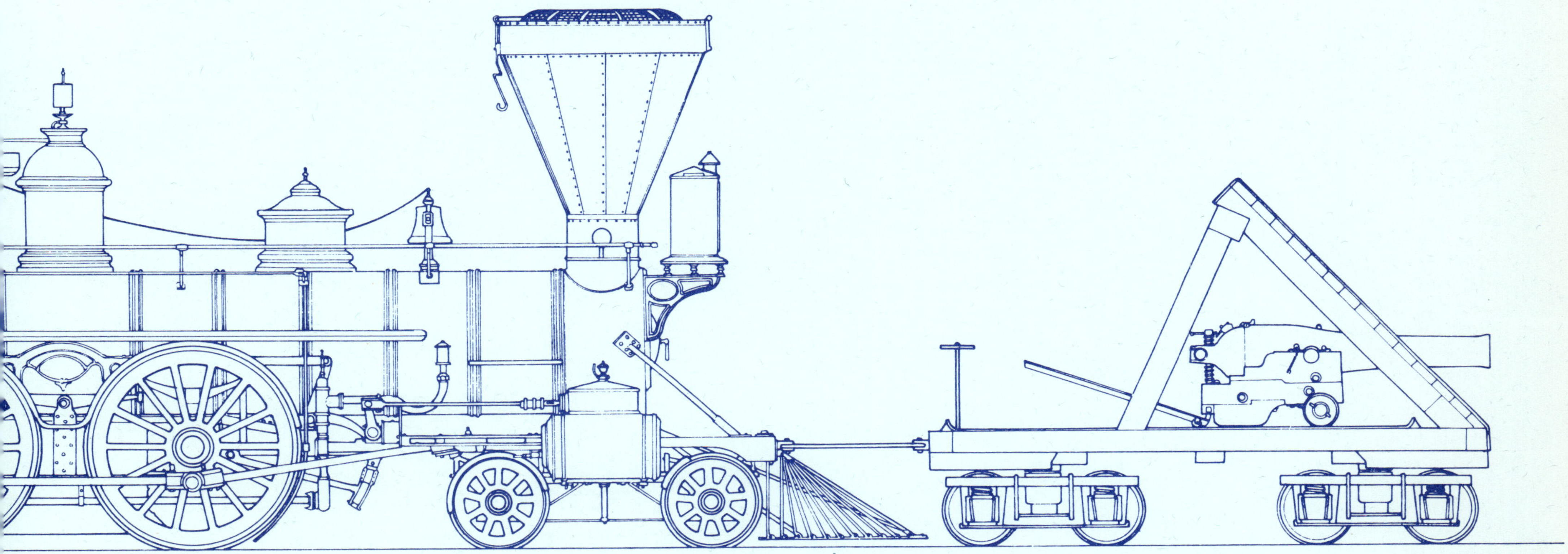

holders to think about and kept up its interest, exhibiting another of the 12cm models at the Paris Exhibition of 1900. Shortly after this they offered a 155mm howitzer on a hydro-pneumatic gun-lift mounting fitted to a suitably strengthened well-base truck. This type of mounting was a standard Schneider coast pattern, which allowed the gun to be lowered behind cover when not in action, and mounting this on a railway truck allowed the gun to be lowered for travelling so as to pass the normal loading gauge easily. The mounting also allowed all-round traverse, being bolted to a turntable unit which was in turn attached to the wagon bed, and stability when firing across the track was ensured by outriggers and rail clamps. This did away with the need to find (or lay) a length of suitably curved track in order to point in the required direction.

Small as the market for these weapons seemed to be, Schneider persevered, and in 1910 sold a number of 20cm howitzers to the Peruvian Government for coastal defense. These were also turntable-mounted, with a shield and firing platform for the detachment, in a well-based double bogie truck with outriggers and rail clamps. It might be explained here that the use of howitzers for coastal defense was probably at its zenith during the first decade of this century, the reason being that these weapons could drop heavy piercing shells down on to the relatively light deck armor of warships, an easier proposition than trying to batter a way in through the side plating. Given this application as feasible, the railway truck was a good method of mounting them; in the first place they usually weighed a good deal less than a gun of comparable caliber; and secondly, the recoil force would be largely downward, rather than across the mounting as would be the case with a low-trajectory gun, and this helped the stability problem. Another point in favor of the railway-mounted howitzer was that this type of weapon had no need to be emplaced on the shore in sight of the target. The static mountings for this class of weapon were usually a short distance inland, concealed from the sea by some terrain feature, so that provided the railway line ran somewhere close to the sea, exposure of the weapon was neither necessary nor desirable.

Schneider's success with the railway gun as a commercial sales item gave them a certain expertise on the subject which was to stand them in good stead in the next few years, and the interest shown in their products began to stimulate designers in other countries, though little in the hardware line was done. But before going further to consider the weapons which proliferated in the 1914–19 period, it might be as well to pause here and, with the railway gun finally established, to consider a few technical points.

The railway gun mounting falls into one of three broad classes—

The Sliding Mounting. In this class the weapon is brought to its firing point and then the frame of the mounting is lowered so that it rests either on the top of the rails or on a prepared base; the wheels are entirely relieved of weight. On firing, that portion of the recoil not absorbed by the normal type of recoil system—if one is fitted—will be translated into a sliding movement of the entire mounting along the rails or along the prepared base. As will be seen later, many sliding mounts have been constructed without the use of a recoil system between gun and mounting, so that the entire recoil force had to be absorbed in moving the mounting. It is as well at this point to remember that recoil is a function of the relationship between shot weight and the weight of the gun and mounting. The interpolation of a recoil system between the two somewhat complicates the arithmetic, but it must be obvious to even a layman that the average railway gun, weighing something of the order of 120 tons, firing a shell weighing about 500 pounds, must have a relatively small recoil. The old soldier's stories about guns which recoiled down the track and into a convenient tunnel to conceal themselves, or recoiled down to the next station and brought back the rations and mail on the return trip, have very little basis in fact.

The Rolling Mounting. Here the weapon goes into action with its weight still on its wheels. The brakes are applied and when the gun is fired, the recoil force not absorbed in the mounting is expended by driving the whole equipment backwards on its wheels, overriding the brakes until they are able to arrest the movement. This type of mounting obviously has to be provided with sufficient wheels to spread the load of the firing shock, otherwise the first round will produce a complete set of flattened wheels.

The Platform Mounting. In this class the weapon is not rigidly aligned with the mounting, as in the two previous classes, but, like the original Schneider designs, is assembled on a turntable or pivot which allows it to be trained independently of the orientation of the mounting. These types are usually fitted with outriggers or stays which are unfolded to give stabilising support behind the gun irrespective of its pointing, and these, together with jacks and rail clamps, ensure that no recoil stress falls on the wheel assemblies. Thus the mounting does not move relative to the track when fired.

It will be appreciated that these types of mounting bring many other factors into play. The sliding and rolling mounts, for example, must be provided with some means of returning the mounting to the firing position after the recoil movement has ceased. Usually the amount of movement is small—three or four feet—and often more than one shot can be fired before the return has to be made, but it has to be done some time, and for this, either a locomotive must be standing by, winches must be laid out and manned, or a hand-powered 'translating gear' can be engaged with one of the mounting axles and the weapon wound back by turning the crank. The locomotive solution is the easiest where a rolling mount is concerned, since all that needs to be done is to release the brakes and push. The sliding mount usually meant that the whole affair had to be jacked back onto its wheels to be moved, though in the lighter equipments powerful winches with adequate ground anchors sometimes allowed the mounting to be dragged back along its platform, sliding it back into place.

Another factor arising from the selection of mounting was the matter of providing the weapon with the necessary amount of traverse to enable it to be pointed at the target. As a general class the railway gun was not expected to cover a wide arc of fire and indulge in opportunity shooting all over the battlefield. It was usually emplaced with a specific target area in mind so that a restricted arc of fire was acceptable. But as anyone who has ever laid

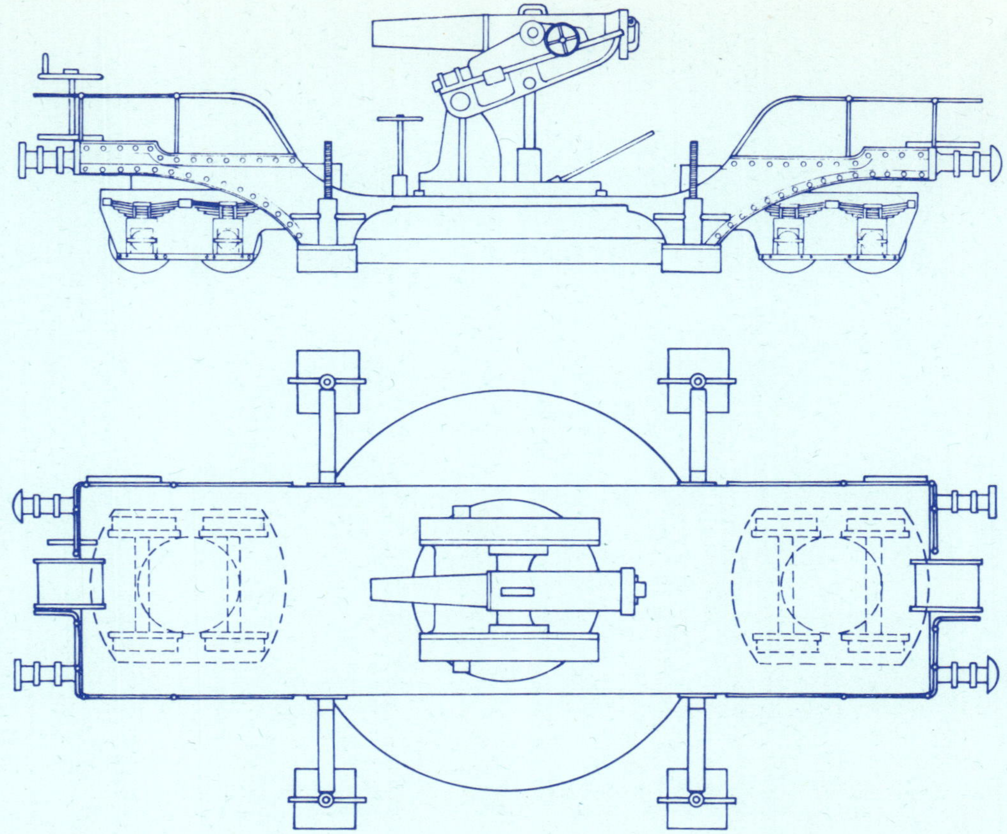

3
The Schneider 155mm howitzer, mounted on a coast pattern hydraulic gun-lift carriage was transplanted onto a railway truck as a mobile coast defense weapon.

4
This Schneider 20cm howitzer was also designed for mobile coast defense and originally sold to Peru. A small number remained in the factory's stock and were turned out hurriedly in late 1914 for the defense of Paris.

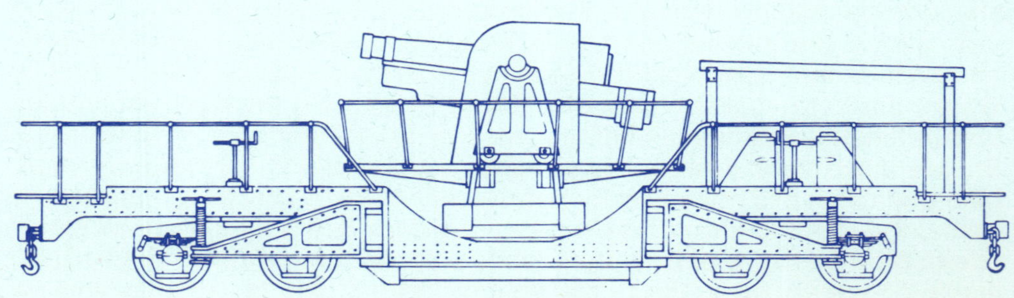

5
Sector traverse

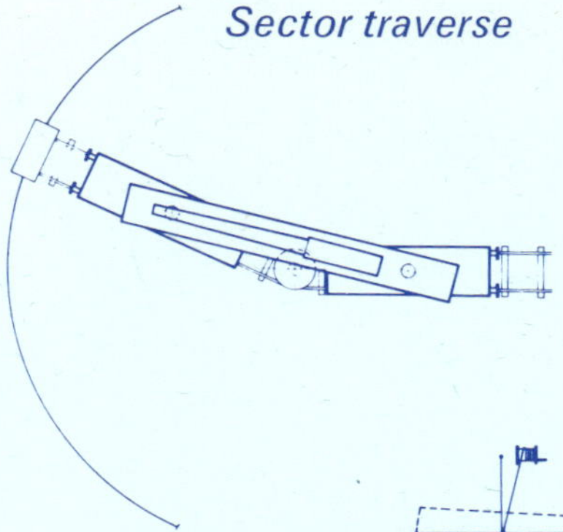

6
Cross-track traverse

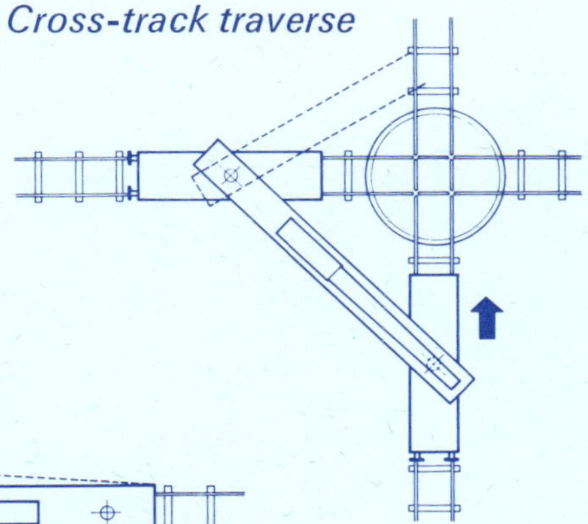

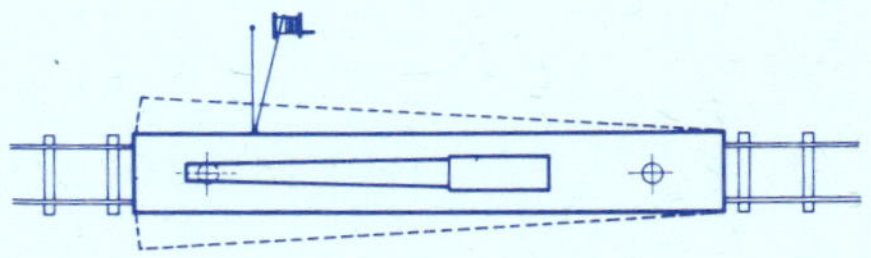

7
Traversing by warping the mounting across the bogies.

a gun or even aimed a rifle can appreciate, some delicacy of control was desirable in order to adjust the fall of shot onto the target. The curved track, as we have already seen, gave the answer for the gun which was rigidly aligned in its mounting; 88 feet of track of 500-foot radius gave about $5\frac{1}{2}$ degrees of traverse, equivalent to a movement of the fall of shot of some 1450 yards at a range of 15,000 yards. But attempting to lay a heavy gun accurately by shouting and arm-waving to a locomotive driver is a time-consuming and frustrating occupation, and it became normal practice to provide some degree of traverse, be it ever so small, on the mounting, to avoid too much fussing back and forth on the track.

One of the first methods used to give fine traverse was to permit a sideways movement of one end of the mounting across the supporting bogie beneath it. This could be done by hand or electrical operation of a screw mechanism, or by simply loosening the attachment and heaving the end of the mounting to one side or another by 'warping winches' set out at right angles. Such a system is obviously restricted by the construction of the mounting, and rarely allows more than one or two degrees of movement.

The other method is to mount the gun and cradle in a 'top carriage' and allow this to be traversed within the main structure of the mounting. Obviously this will not allow very much movement either, but these small switches are all that is required to allow precise laying to be done after approximate positioning of the mounting on the curved track. They will also cater for the misalignment after firing and recoil and so permit re-laying the gun without having to move the mounting back after every shot. Where more traverse is needed, the only answer is to develop a platform mount but, as will become apparent, there is an upper limit of size and power of gun which will suit a platform mount, due to considerations of railway gauge and the recoil forces involved. And if the gun is too big for a platform mounting, then some major engineering is called for.

One understandable question at this point is 'Why?' We have just made the point that the railway gun was usually given a limited arc of fire and usually a specific target area; so why, suddenly, is a larger arc of fire so desirable? The answer lies in the gradual appreciation of the railway gun's utility. In their early days, the specific target theory was acceptable, particularly on the Western Front. But as the guns grew in numbers and in power and range, it became obvious that, given the ability to cover a wide arc, they could command an immense area without having to be moved about. The US 14-inch M1920, for example, had a range of 48,000 yards. If this gun were given an arc of fire of 120 degrees, and assuming it to be emplaced 4000 yards behind the front line, it could drop shells into an area of about 770 square miles. When used as a coast artillery gun, emplaced on the shore and allowed an arc of fire of 180 degrees, it could cover the almost unbelievable area of 1168 square miles of sea. This sort of calculation led to the demand for greater arcs of fire as the state of the art improved, and more particularly when the application of the railway gun to coast defense became popular. Here, in addition to covering a wide arc, the gun had to be able to swing so as to follow a moving target. There is only one worthwhile solution to this; a turntable.

How the turntable was to be married to the weapon was a problem which took several years to solve satisfactorily, but it resolved itself into two lines of approach. One could construct a small but sturdy platform with a roller race, build this into the center of the track, lower the mounting onto it and remove the bogies, as was done with the US 14-inch and the Paris Gun. Or one could install a commercial locomotive turntable at the desired spot and put the gun on it. Since this was hardly practical, a design of portable turntable was eventually developed in Germany which could be laid quickly in any reasonably flat spot and the gun run on to give 360 degrees of traverse.

An ingenious compromise, which gave an arc of about 120 degrees with minimum effort, was the German development of 'cross-track traverse' in which a short cross-track was laid at 90 degrees to the main track, the two being joined by a small turntable. The front bogie of the gun was run onto the turntable and rotated 90 degrees so that it could move along the cross-track, swinging the muzzle end of the equipment about the axis of the rear bogie bolster. An alternative method was to dispense with the turntable, hoist the front end of the mounting into the air, rotate the bogie and drop it in place on the cross-track. It came to the same thing but didn't involve laying the turntable.

Of course, the turntable solution ruled out the use of rolling or sliding mounts. There is little point in going to all the trouble of building a turntable and putting the gun on it if the first shot trundles it backwards off the edge. So turntable mountings demand positive anchorage of the equipment to the table, and a recoil system on the mounting which is capable of absorbing all the firing stress.

A half-way step towards the turntable was the 'sector table', which was a short length of track at the end of a spur line, pivoted to give about 90 degrees of movement. The end of the sector was carried on a wheeled trolley riding on a curved track, and the front bogie of the gun mounting was run onto the sector and clamped in place. The sector could then be swung so that the gun mounting effectively pivoted about the rear bogie. This was not a particularly common system, but was used occasionally by British and French guns.

American Railway Guns

After their innovative steps in the Civil War, the Americans took no further notice of the railway gun until 1917. In April of that year they entered the First World War with a six-inch howitzer as the heaviest piece of mobile artillery, and precious few of them. To provide the expeditionary force with heavy field weapons, numbers of eight-inch and 9.2-inch howitzers were obtained from the British; for heavier weapons, a number of French railway guns were acquired. This gave the troops in France something to train with, and the next step was to develop a manufacturing program to produce some home-brewed designs. Gun barrels were likely to be a production problem, so, since there seemed little likelihood of the German fleet appearing off Long Island, or any other point on the US coast, it was decided to follow the policy already hallowed by the other belligerents, and raid the coast defenses. Some 50 eight-inch guns were removed from coast forts in 1917-18 and assembled to railway mountings in order to provide the AEF with a heavy artillery component. The guns selected were the M1888MI or MII models, and the mounting was based on the standard eight-inch Barbette M1892. A well-base flatcar with two six-wheel bogies formed the basis of the mounting and tubular outriggers were provided to steady the equipment during firing. The Barbette mounting M1918 was assembled to the bed of the car, rotating on a racer ring via a set of roller bearings. An operating platform was fitted to this revolving portion, which carried the gravity loading fixture which became a standard feature of almost all US rail gun designs. The gun was in a ring cradle with two buffer and two recuperator cylinders.

The loading system was simple and effective. A shell table was fitted at the rear of the operating platform. On top of this table was a shell tray inclined towards the gun breech at an angle of about five degrees. A davit on the platform enabled shells to be hoisted up and laid on the shell tray where they were unplugged and fuzed. The gun was depressed to an angle of five degrees depression, the breech opened, and a 'spanning tray' laid between the shell table and the breech. The shell was then pushed by two men from the table, down the spanning tray and into the breech, and with gravity assisting their efforts, by the time the shell reached the chamber it was moving fast enough to coast in and seat firmly in the rifling. The bag charge was then laid on the tray and pushed into the chamber, the tray removed, breech closed, and the gun could then be elevated for firing.

This system did away with the need for a long operating platform, since there was no need to provide room for half-dozen men and a rammer. It also kept the number of men down, since on many heavy guns almost half the detachment have little else to do than help to ram the shells.

By the end of the war only three of these equipments had reached France, but the contract was completed and the weapons retained as mobile coast artillery. Most were stored, though a small number were deployed in positions in the Chesapeake Bay area. A number were sent to Hawaii in 1934, and in 1941 the remainder were recalled to duty and sent to defend the newly-acquired naval bases in the Caribbean. A small number were also given to the Canadian Army for defense of the Pacific Coast in the neighborhood of Vancouver.

Experience had shown that an eight-inch gun was a useful size, firing a reasonable shell to a useful range but with a moderate size and weight of equipment and an all-round fire capability, and it was decided to produce a more up-to-date weapon. A naval eight-inch gun, the Mark VI Mod 3 A2,

The U.S. eight-inch Mark V1 gun preparing for action. The gun has been emplaced and the ammunition handlers are running the 240 pound shells out from the ammunition car to be picked up by the loading gantry on the gun platform. The breech is not yet open, and the spanning tray is folded up.

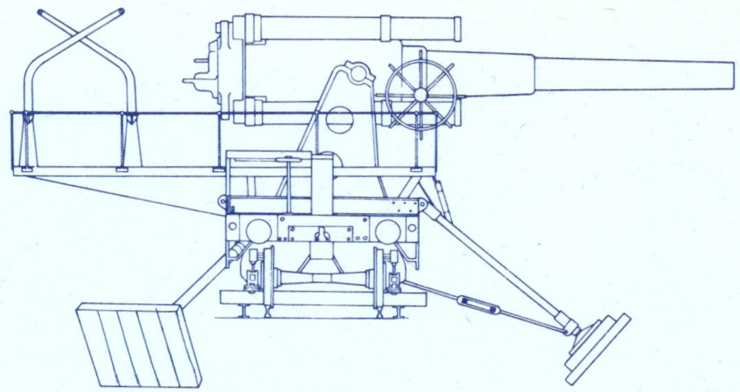

8

Eight-inch Gun M1888MI or MII on Carriage Barbette M1918 on Car M1918MI. The first US eight-inch model; it will be noted that since this was intended for use in Europe it was fitted with European patterns of chain coupling and buffers.

Weight: 77.68 tons (78,930 Kg)

Barrel length: 295.5in (7.505m)

Shell weight: 240 lbs (108.9 Kg)

Maximum range: 23,000 yds (21,030m)

Length between couplers: 45′ 6½″ (13.881m)

Elevation: 42° Traverse: 360°

Velocity: 2200 f/sec (670 m/sec)

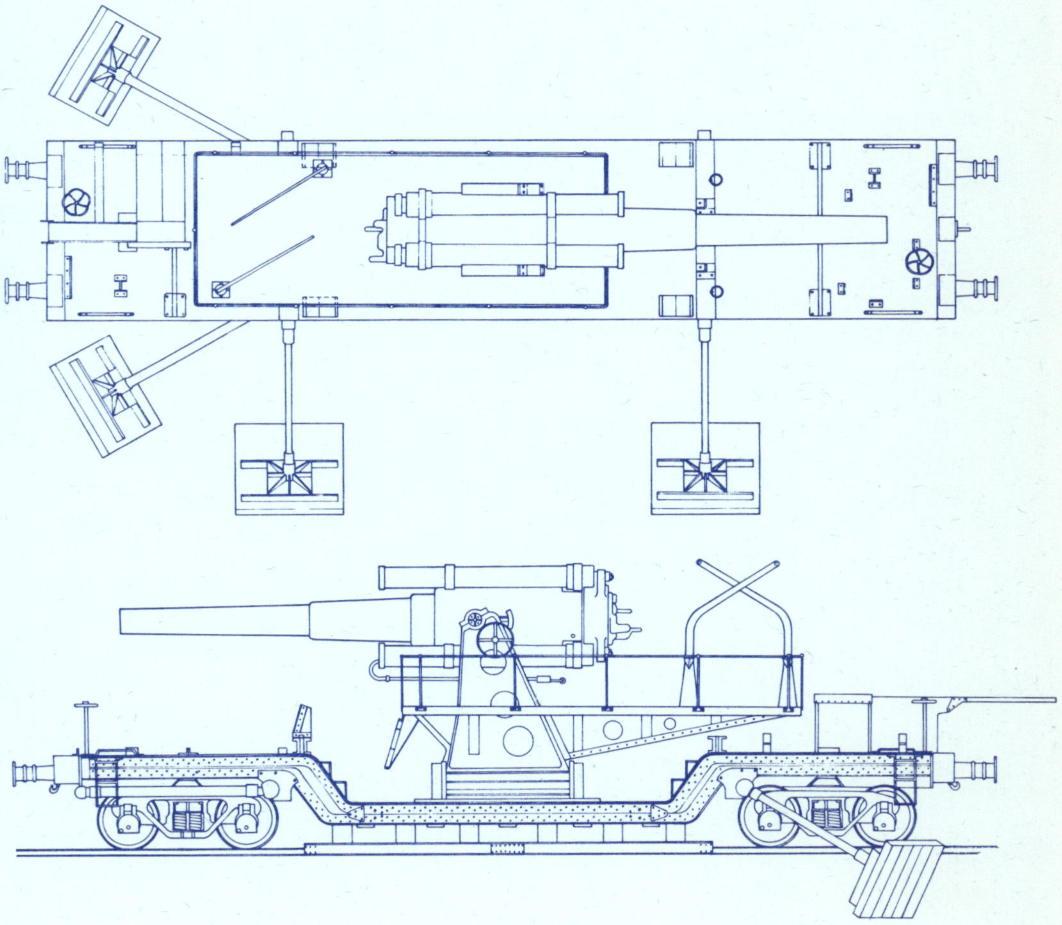

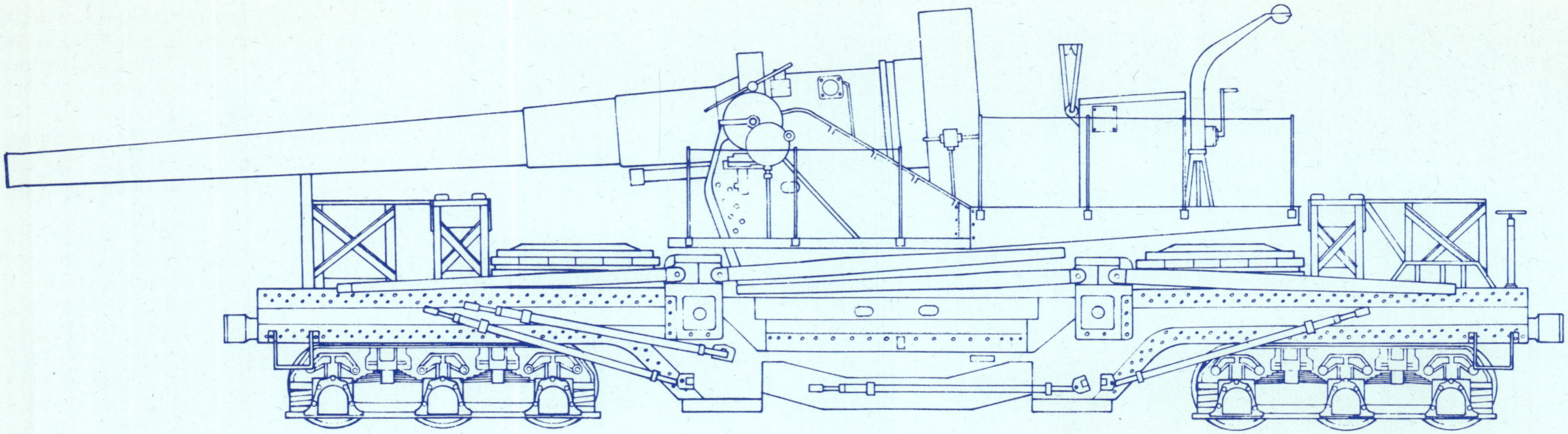

was procured and the chamber modified to take a larger charge. The same type of well-base flatcar with six-wheel bogies was used, but heavier outrigger bars were fitted. The racer and Barbette carriage pattern of mount was retained, though the design was cleaned up and strengthened. A heavy counterweight on the breech end allowed the gun trunnions to be set well back to allow 45 degrees elevation and the ring cradle and recoil system were much improved over the 1918 design.

The guns had become available during the 1920s due to the restrictions on naval armament laid down by the Washington Conference making the guns surplus to navy requirements. At the same time an experimental model of 12-inch howitzer on railway mounting (which never got near service status) was available, and this was used as a basis for the mount design, allowing one of the guns to be assembled and test fired. The design was on low priority, and it was not until 1939 that it was finally approved for production. The first issues were made in 1941 and about 20 were built. Of these, four were allotted to the Pacific coast of the USA, and the remainder to the Atlantic seaboard.

While the eight-inch showed itself to be an effective enough weapon, it was at the low end of the utility scale—70 tons is a lot of gun to drag round for a 240 lb shell—and something more suited to Flanders-style combat was next canvassed. The US Coast Artillery were among the greatest advocates of the coast howitzer—or mortar, as they quite correctly termed them—and large numbers had been installed all over the USA and dependencies. Numbers of these 12-inch mortars were now withdrawn and new carriages and cradles built for them. In their original role they were fitted into a special type of recoil mounting originally designed by Colonel Moncrieff, an officer of the Edinburgh Militia, but this was entirely unsuited to railway mounting. A special ring cradle of cast steel, with recoil and recuperator cylinders above and below, housed the short barrels, and this assembly was trunnioned to a platform mount on a turntable placed on a well-base flatcar, very similar to the eight-inch design.

A ground platform was laid alongside the track and the car moved over the beams of this platform. Using a screw-jack at each corner, the whole mount was then lifted about one inch to allow six wooden beams to be slid in, under the car to rest on the platform. The mounting was then lowered until the whole weight rested on the beams and platform. Eight outriggers were then swung out and their jacks tightened down to stabilise the whole mounting at any angle of traverse; it was also possible to use but two outriggers if firing was restricted to a very small arc about the axis of the track. The whole system was simple and robust, and the weapon could be brought into action in about a half-hour with a well-trained detachment.

About 100 of these were built, though none were available to the army in France before the war ended. In the postwar years they were allotted to coast defense, their original role. In this they were probably more useful on the rail mounting than they had been when mounted statically, since on the rail mount they were capable of firing at any angle of elevation up to 65 degrees, whereas on their original mountings, firing was restricted to elevations between 45 and 70 degrees—hence the term 'mortar': it would have been proper to rechristen them 'howitzer' on the rail mounts.

During the interwar years, most of them were placed in store. Eight went to Hawaii in 1921, and a small number were deployed in the Chesapeake Bay area and around Puget Sound as part of the fixed defenses there. The Second World War did not alter this arrangement, and the stored guns were not taken into use. In 1942 all the weapons were scrapped.

10
The train make-up for a battery of eight-inch Mark VI guns. The gondola cars act as spacers, necessary because of the 7 ′ 2″ overhang of the barrels. The guns are followed by four ammunition cars, two stores cars, a water tank, the fire control car, kitchen car, baggage car, and five coaches for the gun crews.

9

Eight-inch Gun Mk VI Mod3A2 on Mount Railway M1. The improved design, using a converted naval gun barrel.

Weight: 101.3 tons (102,935 Kg)
Barrel length: 369 in (9.872m)
Shell weight: 240 lbs (108.9 Kg)
Maximum range: 35,300 yds (32,280m)
Length between couplers: 42′ 2″ (12.852m)
Elevation: 45° Traverse: 360°
Velocity: 2750 f/sec (838 m/sec)

During the major rearmament of the American coast forts in the 1880s a large number of ten-inch guns had been installed, in both Barbette and disappearing carriage mountings. By 1917 they were considered obsolescent, having been largely superseded as first line armament by the 12-inch gun, so these ten-inch models became the next candidates for conversion to railway mounting. The guns on the disappearing carriages were built with particularly massive trunnions and so the oldest pattern, the Model of 1888, were removed and sent to France to be fitted into sliding mountings of the Schneider type. The finished article was of identical appearance to the French weapons, and a small number were completed before the end of the war and placed in action. Afterwards they were shipped back to the USA, but since the sliding mount was considered to be a retrograde step, excusable only by the circumstances of the time, they were soon declared obsolete and scrapped.

An interesting economy note was that the carriages from which these guns had been taken were also dismantled for scrapping, but the base ring, rollers and racer (the steel turntable upon which the carriage had revolved) of one of them was salvaged and used to form the basis of an experimental ground platform to provide all-round traverse for the 12-inch gun on Batignolles mount. The idea went as far as a prototype, but it was not considered practical and the design was not pursued.

Since the US ten-inch was identical to French designs, we have not included a drawing, but for the sake of completeness the basic data are given here:

Weight: 202.68 tons (205,935kg)
Barrel length: 340in (8.636m)
Shell weight: 575 lbs (261kg)
Maximum range: 20,000 yards (18,285m)
Length: 62 feet 9 inches
Elevation: 54 degrees
Traverse: nil
Velocity: 1975 ft/sec (600m/sec)

With the rapid increase in heavy artillery demanded by the AEF showing no sign of abatement, the forts of the continental US were raided once more and a number of 12-inch guns, Model of 1895, were withdrawn. These were originally on Barbette carriages in their coast defense role, and thus their cradles and recoil systems were available with them. The next problem was what to put them on. The French had, by this time, produced a very sound and simple mounting which was known, from its designer, as the Batignolles Mount. It was a simple box-like assembly carried on bogies at each end. To go into action the structure was positioned on an epi track and cross-ties and girders laid between and alongside the rails. The mounting was then anchored to the bearers, removing the weight from the wheels. With the trucks left in place, no firing stress was taken by them and the girder structure beneath the mounting prevented any hammering of the track on recoil. The original hydro-pneumatic recoil system from the Barbette mount was suitably modified to fit the rail mounting and thus the recoil was absorbed without the mounting moving. Two of the firing platform assemblies, each carried on a special platform car, were provided with each gun so that an alternative firing position could be prepared while the gun was still in action.

Conversion of these weapons began in 1918, and the obvious and simple solution was to ship the guns, with their cradles and recoil systems, to France and there assemble them to their mountings with French assistance. This, of course, took time, and in fact no guns were ready for use until almost the end of the war. As soon as the war ended, the conversion contract, like every weapon program, was severely cut back, and only a small number of weapons were assembled. These were shipped back to the USA in 1919 with the intention of either storing them in their original role as siege weapons or assimilating them into the coast defense system. The unmounted guns were also returned and most of them found their way back into their Barbette mounts and took up their original role. But the Batignolles mount was utterly unsuited to coast defense, having but two degrees of traverse available before the mount had to be jacked up and moved, by means of hand cranks geared to its wheels, to give further deflections. Such a cumbersome system had no utility against moving targets at sea, and the Batignolles mount was relegated entirely to store with its role firmly defined as heavy siege and support firing, the sort of employment which had been their lot in France.

They remained in store until the 1930s, by which time it was realized that, except in the unlikely event of a repetition of the 1914-18 war, there was never likely to be any useful employment for them in the future. They were therefore declared obsolete and scrapped in 1936.

In an endeavor to provide a weapon of considerable range and covering power the next gun to be considered was the 14-inch Model of 1909. This was the gun used to provide the principal armament of the famous 'Concrete Battleship', Fort Drum in Manila Bay, and a reserve barrel was taken as the starting point of the design. At the same time however, the US Navy decided that it really ought to get into the act somehow. Perhaps inspired by the activities of the British Navy and their battery of 15-inch howitzers, with which they added some degree of nautical color to the Western Front, or by the use of battleship guns as railway equipments by other combatants, they also began developing a railway gun using a naval 14-inch gun barrel. By some undisclosed feats of organization and legerdemain they actually managed to get their equipment built, shipped to France and in action by October 1918, and by the end of the war five were in action in support of the US 1st and 2nd Armies in the area of Verdun.

The naval weapon was an unusual design of rolling mount with the breech end of the gun concealed within a coach-roof structure. With no traverse on the carriage it had to be emplaced on a curved track. While it was an interesting design, and a testimony to the navy's acumen and determination, it was not the sort of weapon the army had in mind, and they remained unimpressed. After the war they were retained for a short

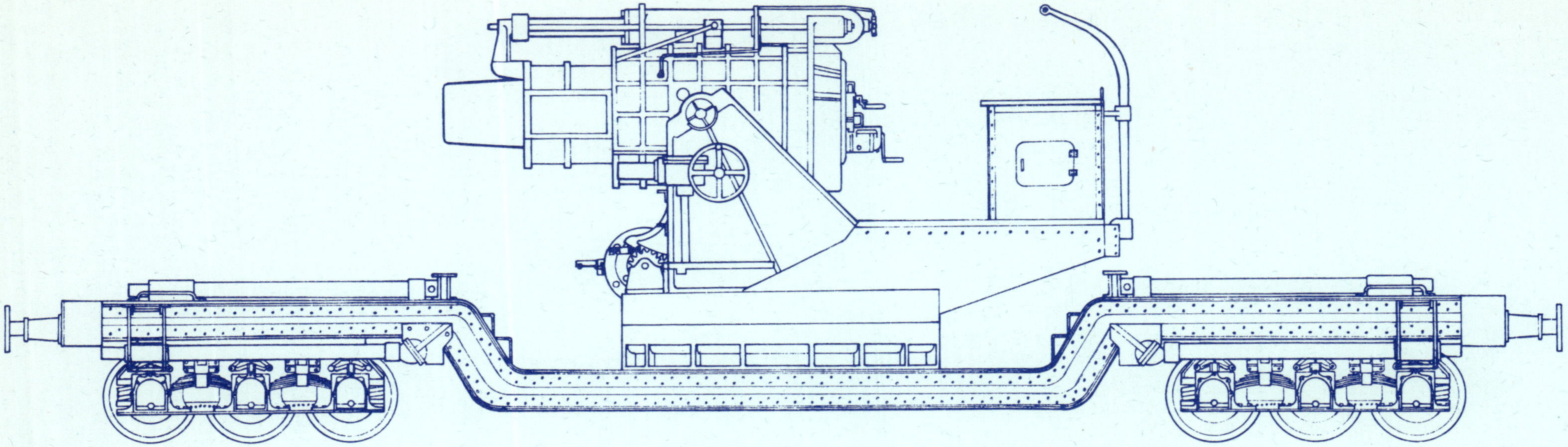

11
12-inch Mortar M1890 on Carriage Mortar 12-inch M1918 on Car Railway Mount M1918MII.
Weight: 78.93 tons (80.195 Kg)

Barrel length: 120 in (3.048m)
Shell weight: 700 lbs (317.5 Kg)

Maximum range: 15,290 yds (13,970m)

Length between couplers: 43′ 5″ (13.233m)
Elevation: 65° Traverse: 360°
Velocity: 1500 f/sec (457 m/sec)

12
12-inch Gun M1895 on Mount Railway Gun M1918 (Batignolles)
It will be noted that this was designed with US standard couplers and not fitted with European pattern buffers.
Weight: 149.5 tons (151,910 Kg)

Barrel length: 443 in (11.252m)
Shell weight: 900 lbs (407.5 Kg)

Maximum range: 28,200 yds (25,785m)

Length between couplers: 66′ $4\frac{1}{8}$″ (20.221m)
Elevation: 38° Traverse: 10°
Velocity: 2600 f/sec (792 m/sec)

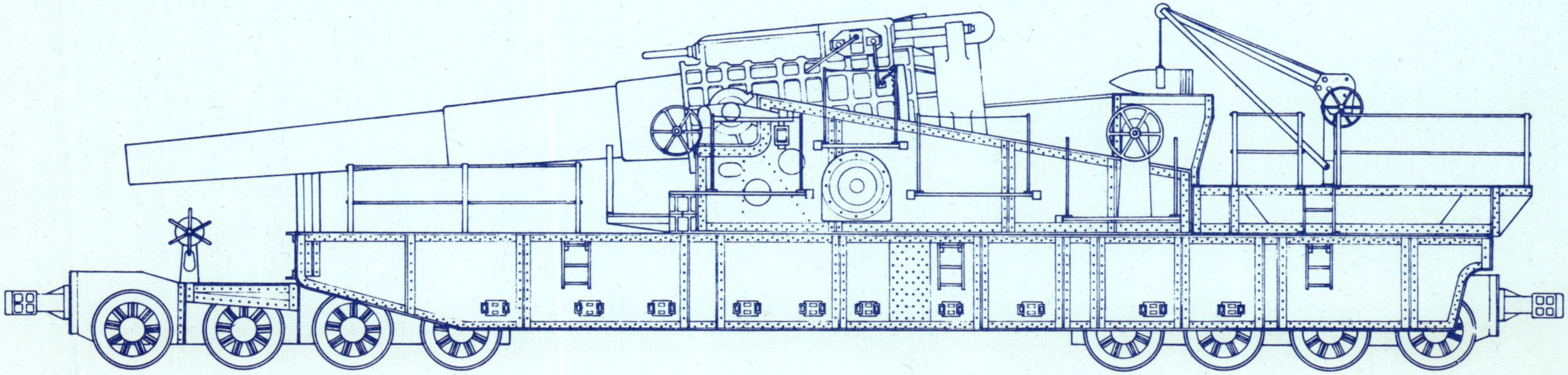

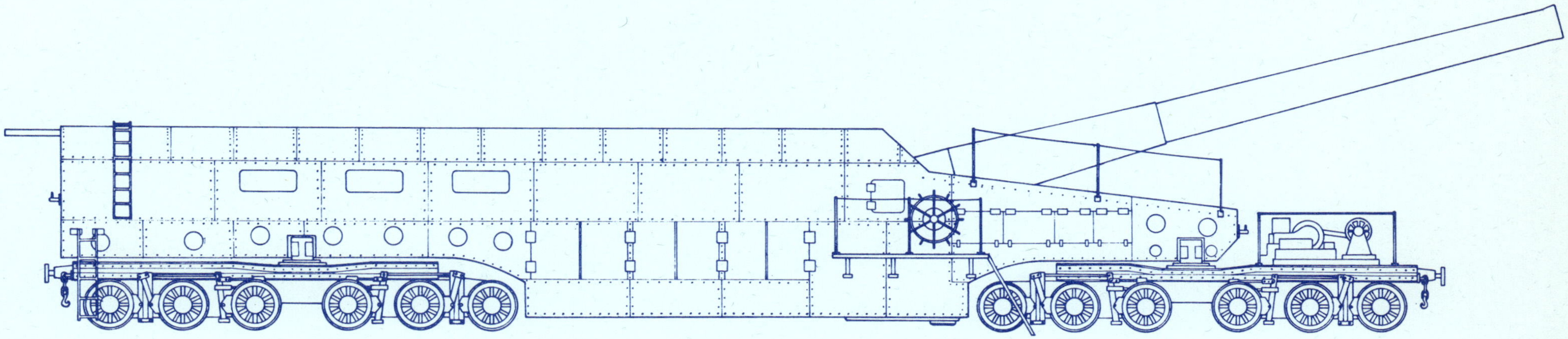

13
The United States Navy design of 14-inch gun. A generator above the front bogie provided power for ammunition handling and elevation, but in the event of failure the ammunition could be manhandled and the large ship-pattern wheels on the side could be used for manual elevation.

14
The United States Army designed their 14-inch Model 'E' based on a coastal gun barrel. While basically a sliding mount it was also given facilities for all-round traverse in the coast defense role.

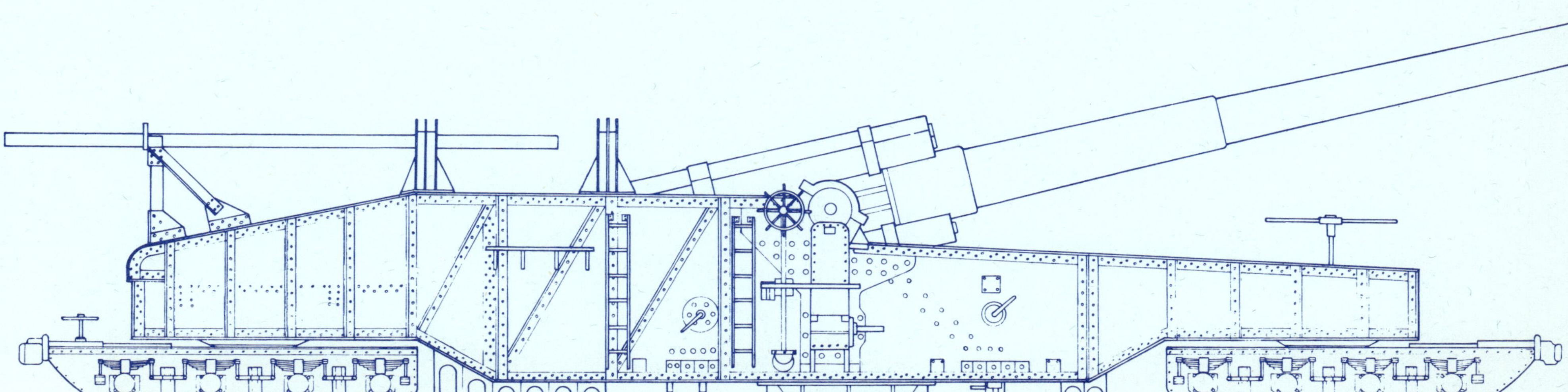

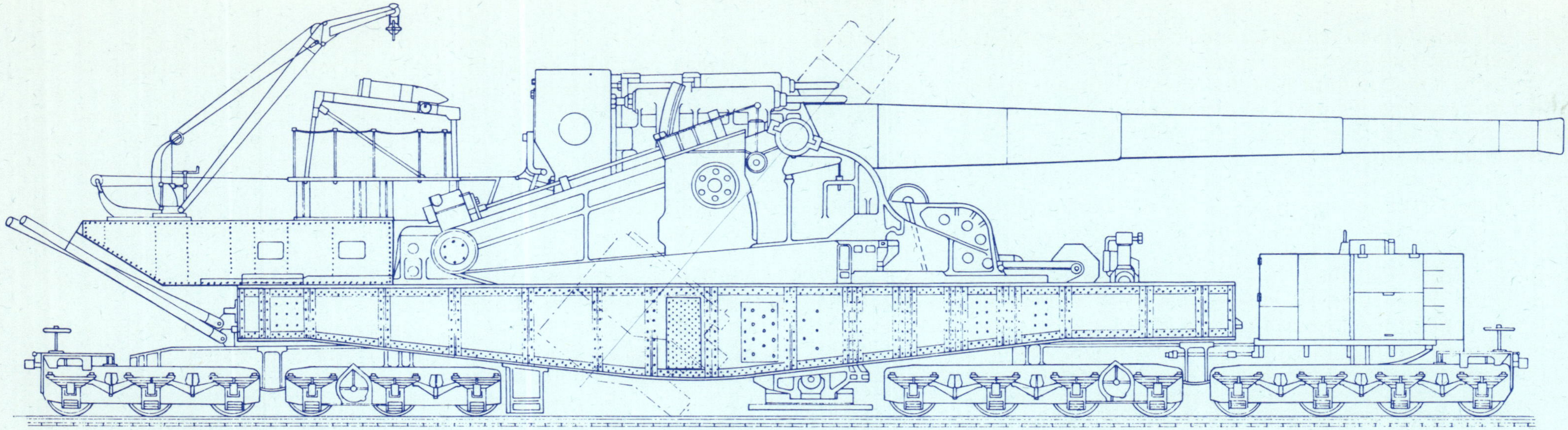

15
14-inch Gun MI920MII or M1920MI or Mk IV Mod 1 on Mount Railway M1920 (Of the guns listed in the official nomenclature above, the Mk IV Mod 1 was the original navy design; the M1920MI was the army modification with the vertical axis canted 16° to fit into the recoil system, and the M1920MII, approved as standard, had the breech axis upright and the recoil system modified to suit.)

Weight: 325.9 tons (323,500 Kg)
Length between couplers: 85′ 6″ (26.061m)
Barrel length: 714 in (18.135m)
Elevation: 50° Traverse: 7° or 360°
Shell weight: 1200 lbs (544.3 Kg)
Velocity: 3000 f/sec (914 m/sec)
Maximum range: 48,220 yds (44,090m)

16
US 16-inch howitzer Model 1920. Originally begun in 1918 in reply to a demand from General Pershing for a heavy weapon with a range of 20,000 yards or more, the development of this calibre terminated with the Armistice, one experimental and not particularly successful model having been built. The idea was revived in the 1920s with the intention of using the 16-inch seacoast howitzer as the basic weapon, but this was little more successful and the equipment was never accepted into service.

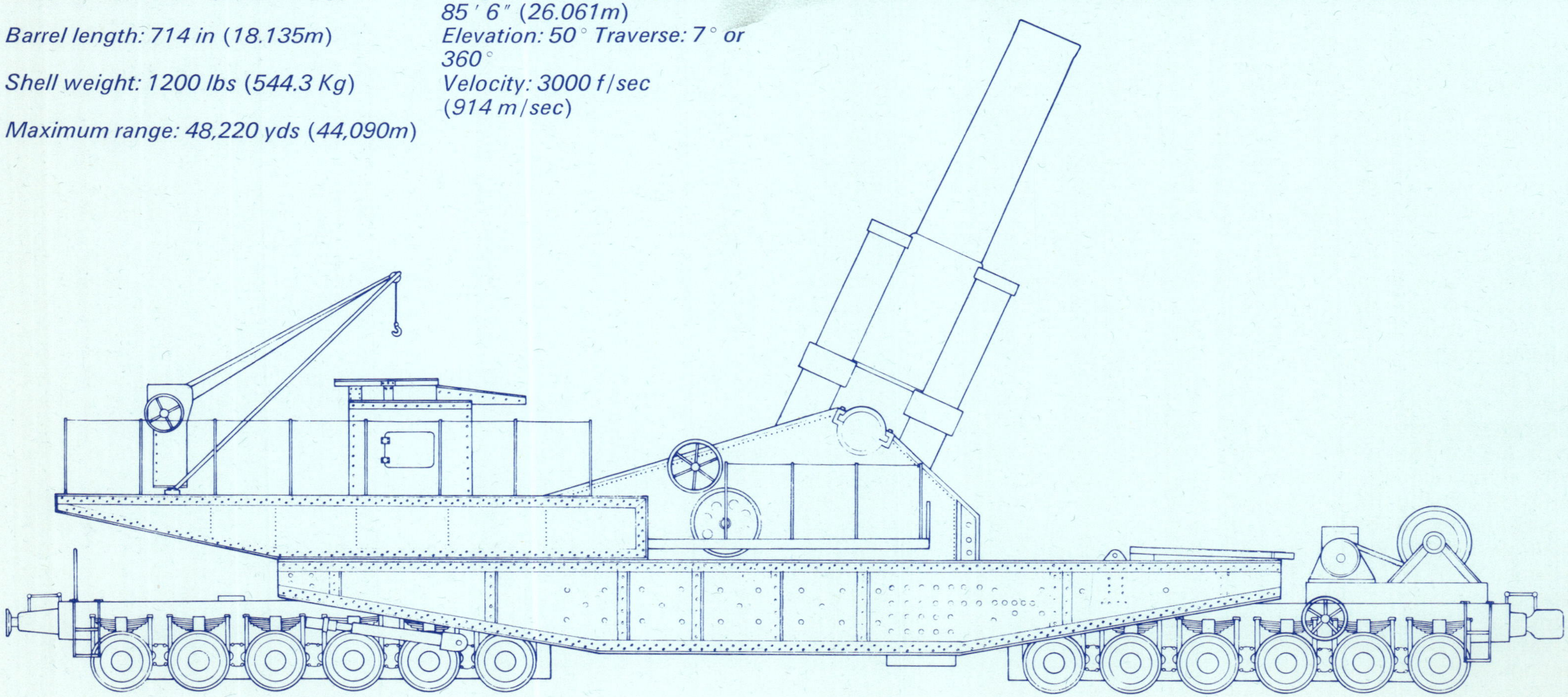

time, but since it had no useful application, being surpassed by the later army designs, it was scrapped in the 1920s.

The army's first attempt was the Model E mounting. This was a simple girder box carrying the gun and recoil system which, as it stood, could be jacked down onto a prepared bed and fired as a sliding mounting. A considerable improvement was then made in the design by developing a portable firing platform consisting of two semi-circular steel castings which were hinged together and could be folded and carried on an ordinary flat car in the gun train. A heavy crane car was provided in the gun train for handling this platform. After selecting a site and leveling the ground, a section of track was removed and the platform lowered into the gap and opened out. Steel stakes were then hammered into the ground around the circumference of the platform to hold it securely in place. The crane was then removed and the gun mount run forward across the platform, jacked up, the bogies removed, the mount lowered until four rollers on the underside rested on a roller track in the platform and a pivot socket fitted over a pivot in the platform centre. The platform casting had a rack machined on the inner face of the roller path, and a pinion on the mounting was engaged with this so that by using hand cranks to drive the pinion the gun could be traversed through 360 degrees. Once pointed in the right direction, circular steel pads were laid beneath the rear end of the mounting and screw jacks used to take the weight and firing stresses. The whole operation of laying the platform and emplacing the gun took about eight hours.

Once the use of railway guns for coast defense had been officially approved the designers perfected a technique of preparing a concrete base for this firing platform. This could be prepared at leisure in any likely site and when needed the steel platform could be dropped on and bolted down and the gun emplaced in little more than an hour.

The development of the Model E had begun in mid-1916 and was not completed until late in 1918, and good as it was, it was still not good enough in the eyes of the army, particularly since now that the stress of war had been removed, the best possible design could be developed without worrying about time. So a final design was produced, an amalgam of the naval gun barrel, which was a more modern and powerful design than the army model, and an improved Model E mounting.

With this size of weapon the designers came up against the problem of the railway loading gauge. If the gun was to have a reasonable amount of elevation—and it must, since otherwise there was no justification for adopting it—then the trunnions had to be set high from the roadbed in order to allow the breech end to recoil without striking the track. But building the trunnions at this height meant that the gun would be too high to pass beneath bridges and other railway structures. The problem was solved very neatly. Firstly the recoil system of the 16-inch Barbette gun carriage was adapted to the cradle, reducing the length of the recoil stroke from the Model E's 60 inches to 35 inches. Then the mounting was built with a top carriage, a heavy yoke hinged to the main structure at its rear end and carrying the gun and cradle at its front. The front of the yoke was connected to the main girders of the mount by a pair of massive screwed rods passing through threaded collars connected to the mounting. By rotating these collars, the screwed rods were lowered and the entire yoke and gun assembly dropped into the body of the mounting until the whole structure was low enough to pass any railway gauge. In this position the yoke rested on a heavy steel transom connecting the side girders and the lifting mechanism was relieved of any weight.

To prepare to fire, the collars were revolved to lift the screwed rods, yoke and gun clear of the upper section of the mounting side girders, when a heavy steel beam known as the 'firing support' was inserted between the yoke and the mounting and the yoke lowered until the weight was taken on the firing support. Contact between the yoke and the firing support was through two traversing rollers. The yoke was pivoted in the vertical plane as well as the horizontal, and could be traversed a small amount across the firing support.

For firing the gun was run into place on a curved track and a number of steel beams laid alongside the rails. The mount could then be lowered on to these beams and the wheels relieved of weight. Outriggers were swung out at front and rear to act as recoil spades, and the result was virtually a non-sliding variation of the sliding mounting.

But this equipment had been designed with seacoast defense firmly in mind from the start, and in this role it was emplaced over a prepared concrete and steel firing platform based on the Model E but considerably simplified. It was a 42-foot diameter concrete slab with a heavy steel baseplate bolted down in the centre and a 15-inch wide racer ring around the edge. The central baseplate had two running rails cast into it, and to bring the gun into action bridging rails were laid across the concrete to allow the mounting to be run across until it was centered over the baseplate. It was then lowered until a pintle plate on the underside of the mounting could be bolted to the baseplate. The bogies were then removed and a traversing unit with rollers was lowered to rest on the racer ring and support the rear end of the mounting. The whole operation took about two and a half hours.

Four of these weapons were built in the late 1920s. The first went to Fort MacArthur, California, followed by two to the Panama Canal Zone, where numerous platforms were prepared for them, and the last was delivered to Fort MacArthur in July 1929. It is said that there were discussions in 1944 over the possibility of sending these guns to Europe in order to provide heavy support for the assault on Germany. At the same time the British Army were contemplating using their 18-inch railway howitzers for the same task, but in the event the speed of the Allied advance, together with the massive air power available, rendered them unnecessary and they remained at their locations until the final dissolution of US Coast Artillery in 1949.

The last American railway gun design was the 16-inch howitzer. This weapon was originally developed as a coast defense piece mounted in a concrete emplacement, and was the last of the high-angle coast weapons. It was introduced in 1920, was declared a 'limited standard' weapon, and only four were ever installed, since the day of the coast howitzer was almost over. But it looked like being a potentially useful siege equipment and during the development stage of the weapon, a design of railway mounting was also put in hand. Known as the Model 1918, it was a rolling mounting with top carriage traverse. It was later improved into the Model 1920, but after mature consideration, the project was dropped. It was unusable as a coast defense weapon on the rail mounting, and the only projectile for it was a deck-piercing shell hardly suited to land warfare, which meant the design and provision of a complete outfit of special ammunition. These objections, and the inevitable shrinkage of funds in the postwar years were sufficient to put an end to the 16-inch howitzer.

French Railway Guns

As we have seen, the Schneider Company of France had been active in the design and production of railway guns in the early years of the century, but the French Army remained unimpressed by these weapons and no railway gun was ever taken into official use. The French Army had spent the years since the 1870 war in developing a philosophy of attack and rapid movement, and the railway gun, indeed any heavy gun, played no part in it. The artillery of the French Army rested on a firm base of 75mm Field Guns Model of 1897, which were among the best of their kind in 1914, and the remainder of the armament was a collection of lighter weapons for horse and mountain artillery and some 12cm and 15.5cm guns and howitzers to provide a leavening of heavy support. The 15.5cm howitzers were available on a scale of rather less than four howitzers per army corps. It was contended that the rapid fire of the 75s would make up in quantity what was lacking in shell weight.

The events of 1914 soon showed that this sort of mixture was the wrong one for the circumstances. Heavy artillery had to be found, and quickly. By raiding the stocks of reserve naval guns, removing guns from fortresses distant from the fighting and milking coast defenses, numbers of guns were made available, and the Schneider company were asked to provide suitable mountings to allow this collection to be put to use at the front. The drawback was that these guns were originally designed to be installed in coastal and naval carriages in which the recoil system formed part of the mounting. The guns were fitted with trunnions and rested in top carriages which recoiled on a slide, the recoil system being interposed between the top carriage and the slide. This meant that there were no suitable cradles and recoil mechanisms which could be adapted to mobile carriages. The manufacture of such items would have taken years, as well as involving the removal of the gun trunnions. The solution finally adopted was of startling simplicity, and was unlikely to have been given two minutes hearing in peacetime. The guns were adapted to railway mounts with no form of recoil control whatever.

The design perfected by Schneider was a simple steel box with trunnion bearings into which the gun was secured. The box was then supported at the ends by sufficient wheels to take the traveling weight, and pulled into its selected position. Since the front line by this time was relatively static, the time required to prepare the site was of little importance. Having laid track to the gun position, the track bed was reinforced by four rows of ten-inch I-section girders laid parallel to the track. The gun mounting was moved over this platform and cross-beams beneath the mounting lowered by screw jacks until the weight was taken from the wheels and completely transferred to the I-beams. When the gun was fired the recoil stress passed via the trunnions to the side girders of the mounting, and moved the whole equipment back, the recoil force being absorbed by the friction between the mounting and the platform. The amount of rearward movement varied from one to four feet, depending on the size of the gun and the amount of elevation. After the recoil movement stopped, the screw jacks were operated and placed the weight back on the wheels, and a hand crank mechanism engaged with a chain drive to one of the axles was operated to wind the gun back to the firing position. There it was jacked up again and the whole cycle repeated. It was scarcely the fastest gun on the Western Front but it worked, it was simple, and it could be manufactured quite quickly, which was what mattered most at the time.

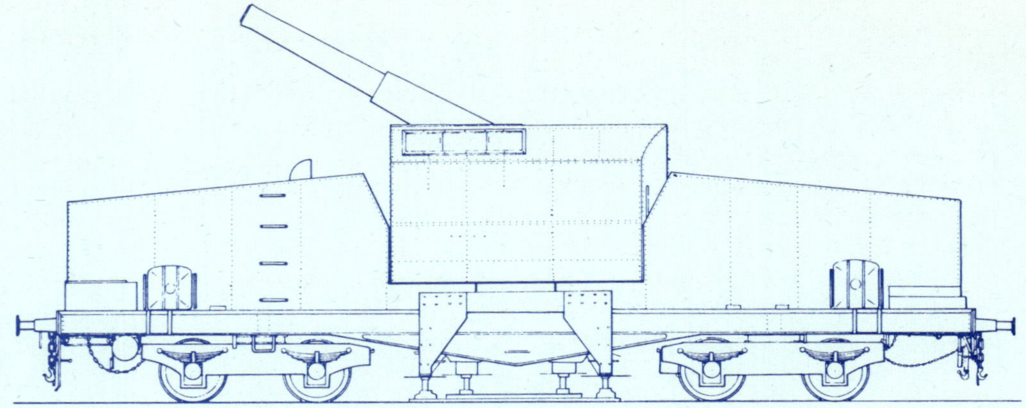

17

Materiel de 194 Mle 70/94 sur affut-truc tous azimut.

Weight: 63.97 tons (65,000 Kg)	*Length: 48′ 2″ (14.68m)*
Barrel length: 231.75 in (5.886m)	*Elevation: 40° Traverse: 360°*
Shell weight: 18.28 lbs (83 Kg)	*Velocity: 2093 f/sec (638 m/sec)*

Maximum range: 20,010 yds (18,300m)

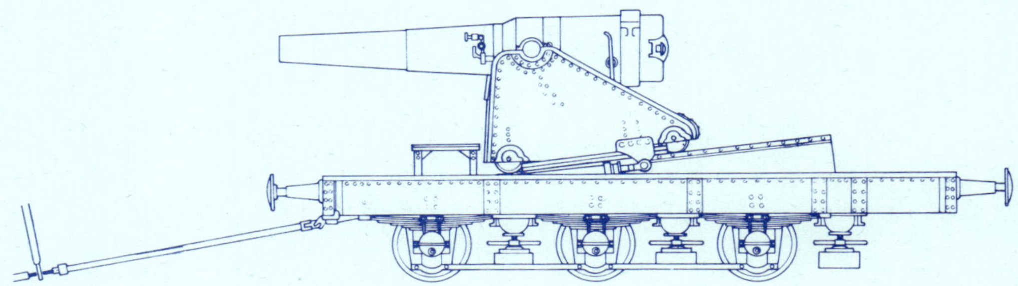

18

194mm Mle 1875. While virtually the same gun as in drawing 16, the mounting is a hurried wartime design using a ground anchor and holdfast at the front. The hydraulic buffer and sloping recoil slide are typical late 1880s coast artillery and naval pattern.

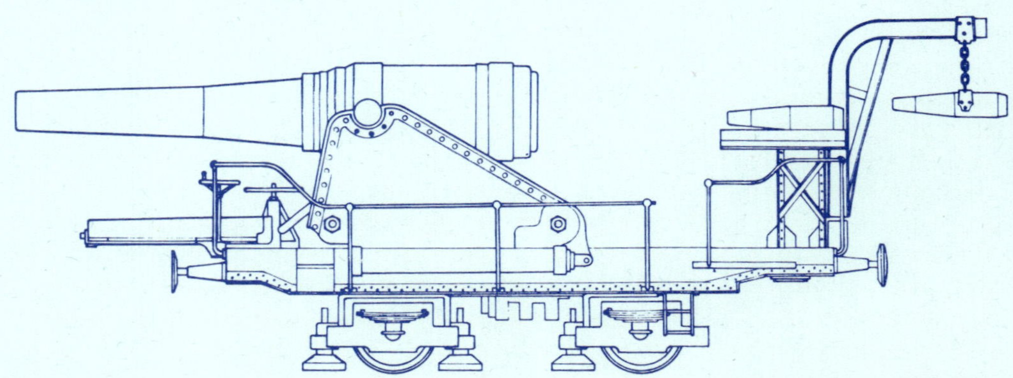

19

Another hurried design was this 240mm Mle 1876, using a similar type of ex-coast carriage but with a flat recoil slide and an improved pattern of hydraulic buffer which incorporated springs to push the gun back into the firing position.

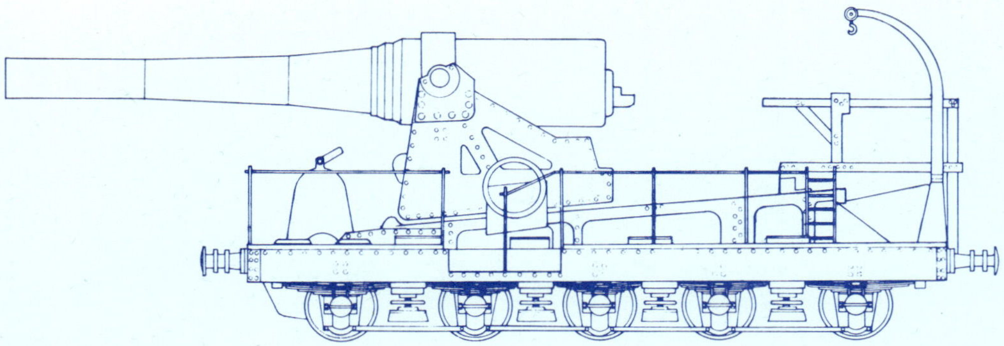

20
Canon de 140 sur affut-truc Mle 1884. An ex-naval inclined slide mounting adapted to rail use in the early weeks of 1915, using hydraulics to damp out the recoil and gravity to return the gun.

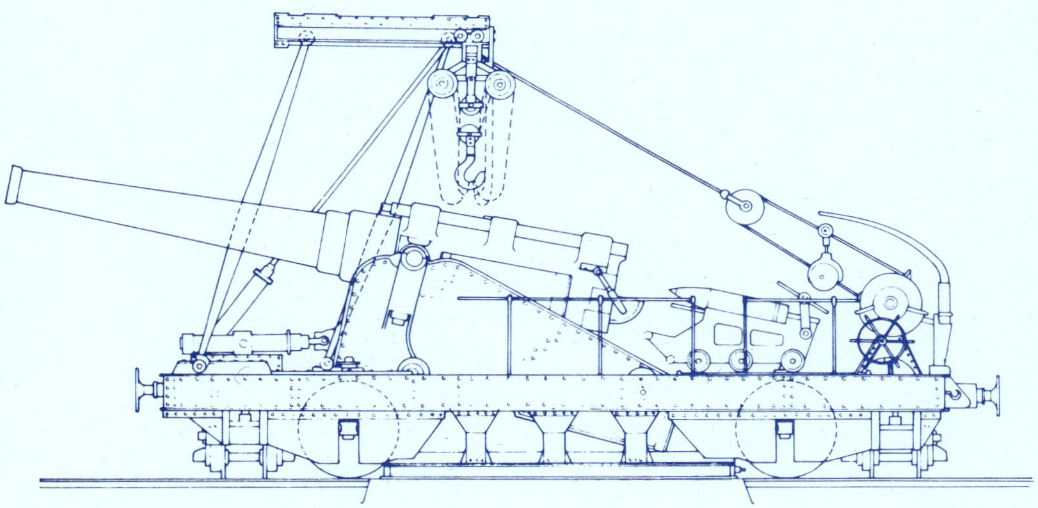

21
Canon de 240 sur affut-truc Mle 1903. This is equipped with an A-Frame for moving the gun on and off its transport truck, as shown in the photographs of the 293mm howitzer.

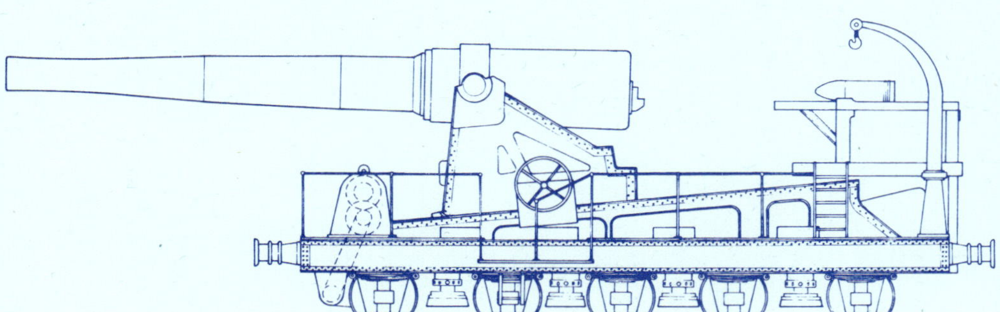

22
Canon de 240 sur affut-truc Mle 70–81. Another ex-coast or naval sliding mounting hurriedly adapted. Like most of the type, a shell-hoisting davit was supplied but little else in the way of mechanical assistance.

The only drawback to this design was that after a few hundred rounds had been fired, the enormous blows of the recoil force would eventually have their effect on the side girders, and these would begin to crack and break away at the trunnion bearings. But by that time the gun barrel was usually worn to its limit of accuracy, so the whole weapon would be withdrawn and scrapped. Apart from that they were effective and popular, since they demanded little in the way of maintenance or overhaul—no recoil system meant no filling and purging, no daily checks, no glands to repack. Probably the only men who didn't like them were the *poilus* who manned the crank and wound them back into place after every shot.

In addition to these *'glissement'* mountings, a number of other odd patterns appeared in 1915, most of which were either weapons which Schneider and other companies had built in prewar years around standard French guns in their attempts to interest customers, or were improvisations of one sort or another to make use of smaller weapons. After the emergency provision of elderly guns there was a more deliberate period of design which produced one or two platform mounts for small guns, followed by the Batignolles mount already mentioned as being used by the Americans for their 12-inch gun. This was considered to be the best limited-traverse mounting produced during the war due to its simplicity and robustness. Special steel cross-ties with integral spades were inserted between the normal track sleepers, and two heavy girders were bolted onto the ends of the cross-ties, parallel with the track. These girders were provided with 12 sets of flanges which matched similar sets on the bottom of the mounting. Once the basic platform was laid, the mounting was run into place and by means of a hand crank geared to one axle was precisely positioned so that the sets of flanges matched. The mount was then locked to the girders by driving in wedges between each set of flanges until the weight of the mounting was directly on the base and not on the mounting wheels. The interlocking of flanges and the integral spades of the cross-ties ensured that both the downward and rearward thrusts due to firing were carried into the ground and the mounting never moved. Some Batignolles mounts were built using guns with cradles and recoil systems, but a number were also built on the Schneider system with no recoil mechanism interposed between gun and mounting.

The last French railway gun to be built during the war was, frankly, a waste of effort. After the Germans had battered Liège, Maubeuge, Namur and Verdun with their 42cm howitzers, public outcry in France demanded that to restore French prestige a bigger and better weapon had to be produced. The designers got to work and came up with a 52cm (20.5-inch) howitzer. Developing a road-mobile mounting for this monster would have taken far too long, so a railway mounting was selected as the quick solution. It was first intended to take the quickest way out and use a Schneider non-recoil sliding mounting, but the recoil energy was too great to be comfortably accommodated, and it was necessary to design a suitable recoil system. Two of these weapons were built and duly paraded for visiting politicians, but little other or warlike use was made of them.

The number of different designs of French rail guns, due to the hurried construction of emergency equipments in 1915, is legion, and space cannot be spared to cover each model in detail. Since, in many cases, the same mounting was used for a variety of barrels, a description of one will have to suffice for all, though the various weapons mounted will be tabulated.

Among the earliest designs were the 164mm guns; these were naval weapons and were fitted to simple platform mountings on a straight-back

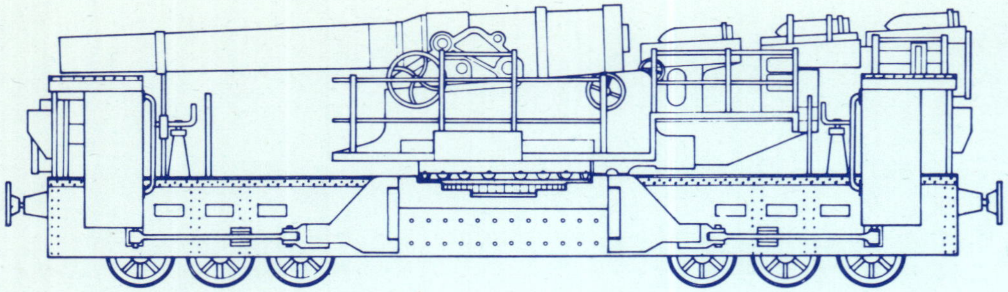

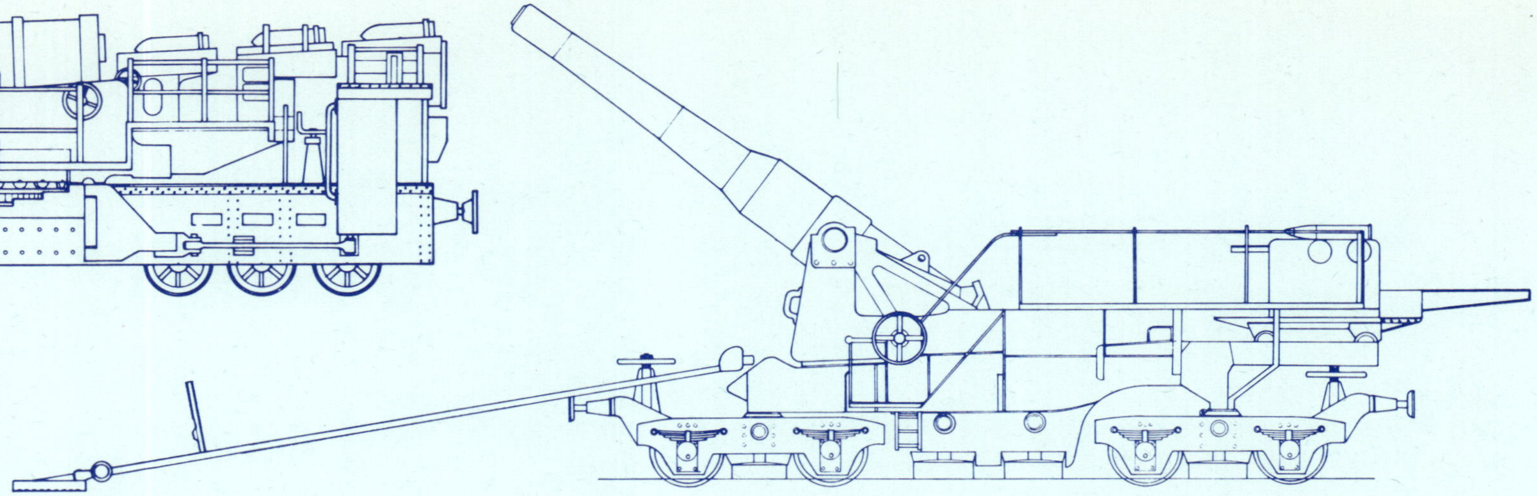

23
Canon de 240 sur affut-truc Mle 1893. Although jacked for firing, this weapon was also secured by the use of a stout ground anchor and holdfast.

24
Canon de 240 sur affut-truc Mle 1884
Weight: 88.58 tons (90,000 Kg) *Length: 42′ 0″ (12.884m)*
Barrel length: 263.75 in (6.700m) *Elevation: 38° Traverse: 360°*
Shell weight: 350.2 lbs (159 Kg) *Velocity: 1885 f/sec (575 m/sec)*
Maximum range: 19,025 yds (17,400m)

25
Canon de 240 sur affut-truc Mle 1893–96
Weight: 138.77 tons (141,000 Kg) *Length: 64′ 0″ (19.5m)*
Barrel length: 395.87 in (10.055m) *Elevation: 29° firing across track, 35° firing along track. Traverse: 360°*
Shell weight: 356.8 lbs (162 Kg) *Velocity: 2755 f/sec (840 m/sec)*
Maximum range: 24,825 yds (22,700m)

truck. The mounting was shielded and armored magazines for ready-use ammunition were provided at each end of the truck. The mounting itself was an ex-naval 'broadside' mounting with top carriage recoil. For firing, four screw jacks were positioned under specially reinforced mounting points and tightened down onto pads placed alongside the track, after which the gun could be traversed through 360 degrees and fired at any bearing. This gun, and a 194mm of similar appearance, were not normally used as railway artillery of position on the Western Front but formed part of special armored trains for coastal defense against the possibility of a German attack in an attempt to outflank the seaward edge of the front line.

An elderly 194mm howitzer was also railway mounted in very limited numbers early in 1915. This, the Model of 1875, was an ex-fortress weapon lifted bodily with its slide mounting and dropped on to a three-axle flat wagon. The wagon was suitably reinforced and provided with screw jacks

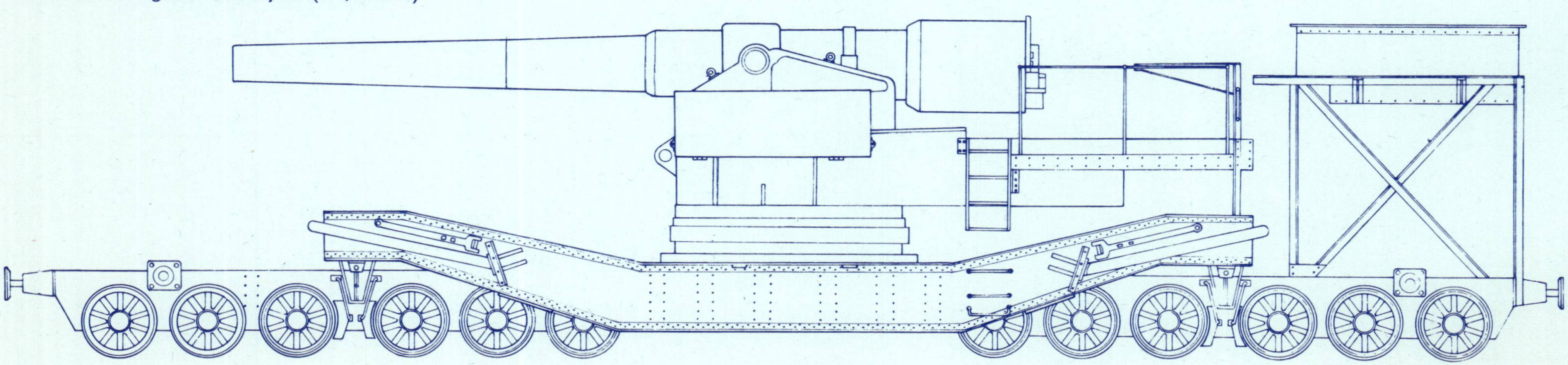

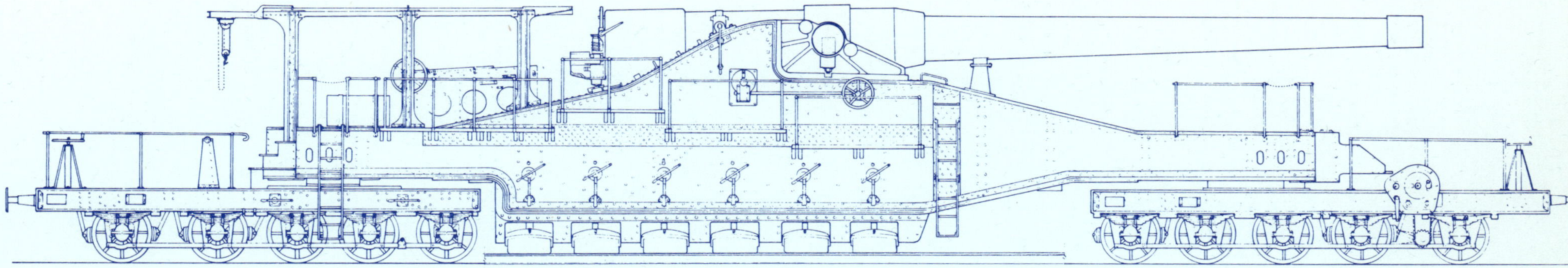

26
Materiel de 274 Mle 87 ou 93 sur affut-truc à glissement
A typical example of the Schneider sliding mount, showing the heavily reinforced trunnion bearings, the jacks to lower the mounting blocks and an overhead rail for supplying ammunition to the loading table.

Weight: 149.6 tons (152,000 Kg)
Barrel length: 503.95 in (12.800m)
Shell weight: 522 lbs (237 Kg)
Length: 85′ 0″ (25.9m)
Elevation: 40° Traverse: Nil
Velocity: 2485 f/sec (758 m/sec)
Maximum range: 28,875 yds (26,400m)

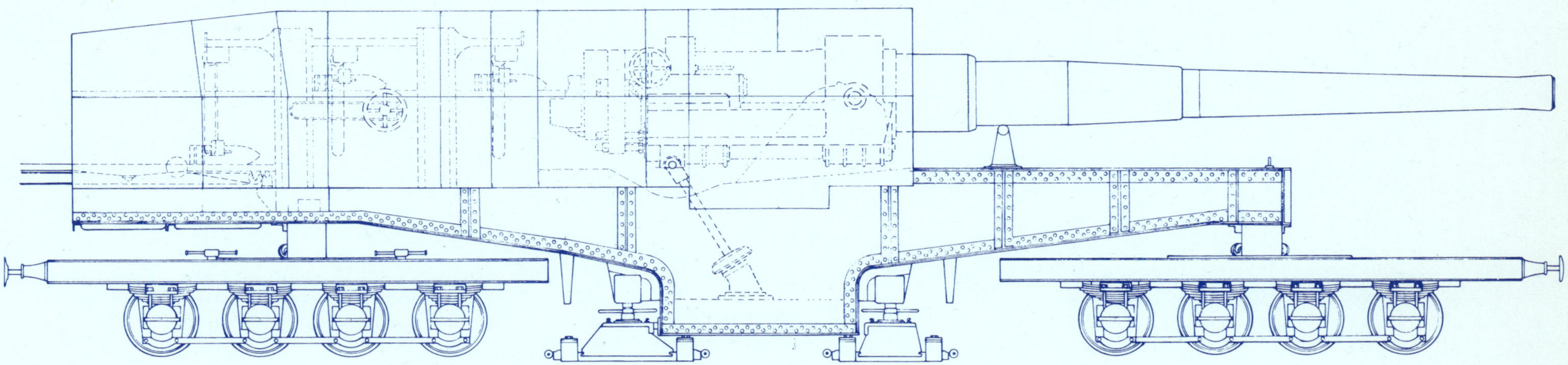

27
Another version of the 274mm Gun Mle 1893 with a covered gun house. Shown dotted are the ammunition hoisting and handling details, the elevating screw and other interior fittings.

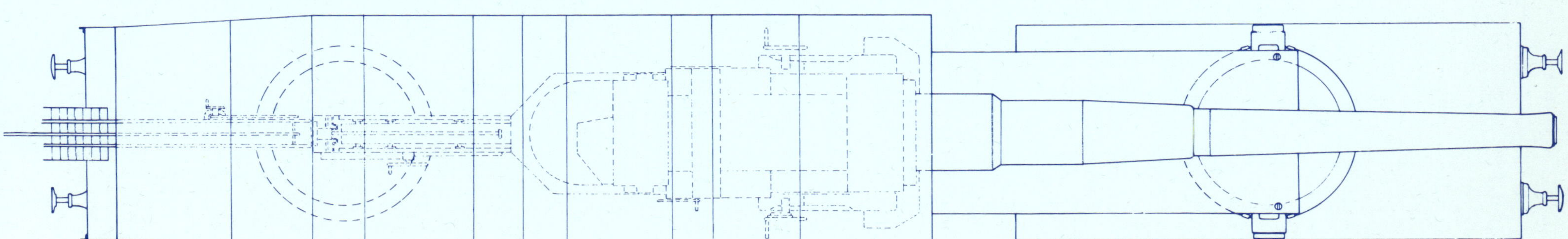

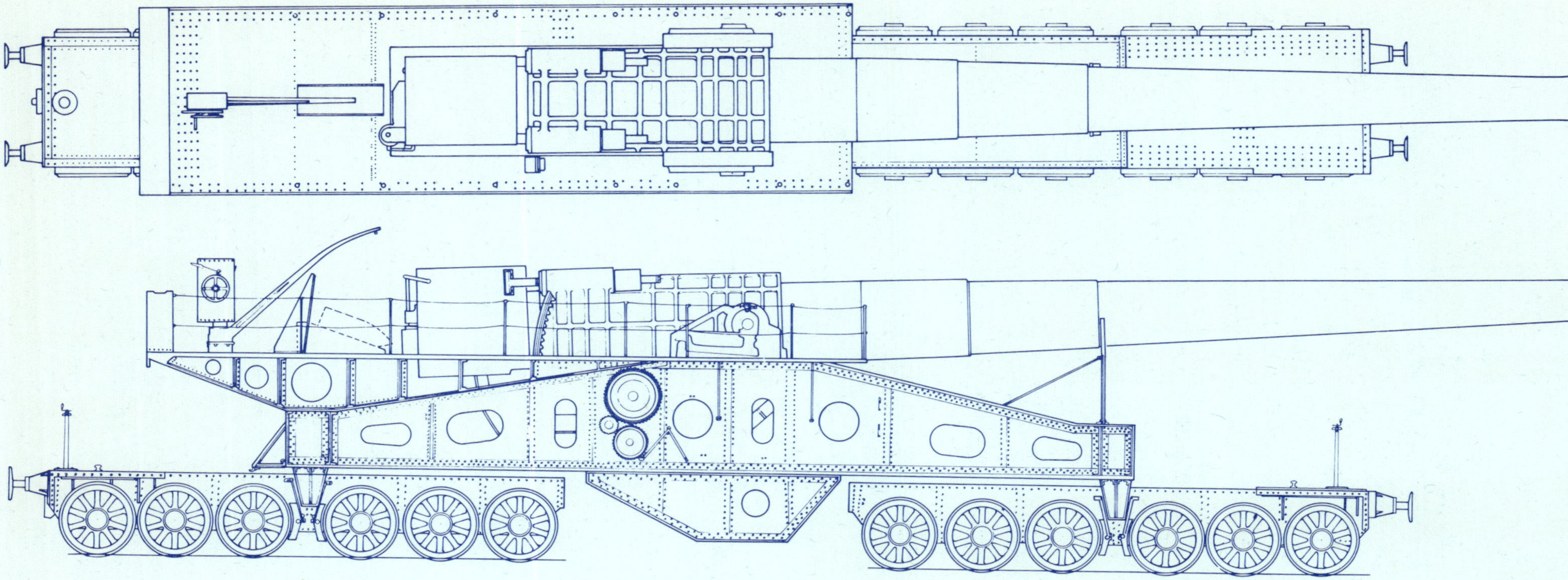

28
Matériel de 340 Mle 1912 sur affut-truc à berceau
The high-powered version of the Mle 1912 needed a cradle and recoil system to absorb the additional recoil stresses.

Weight: 265,5 tons (270,000 Kg)
Barrel length: 634.4 in (16.114m)
Shell weight: 947 lbs (430 Kg)
Length: 110 ′ 0.86″ (33.728m)
Elevation: 37° Traverse: Nil
Velocity: 3040 f/sec (927 m/sec)
Maximum range: 41,120 yds (37,600m)

29
The 293mm howitzer Mle 1903. Due to low stresses this could be mounted on a quite lightweight truck, with the recoil force largely absorbed by a hydro-pneumatic system in the large cradle under the barrel. The mounting of this weapon was done by A-Frame, as shown on the adjacent photographs.

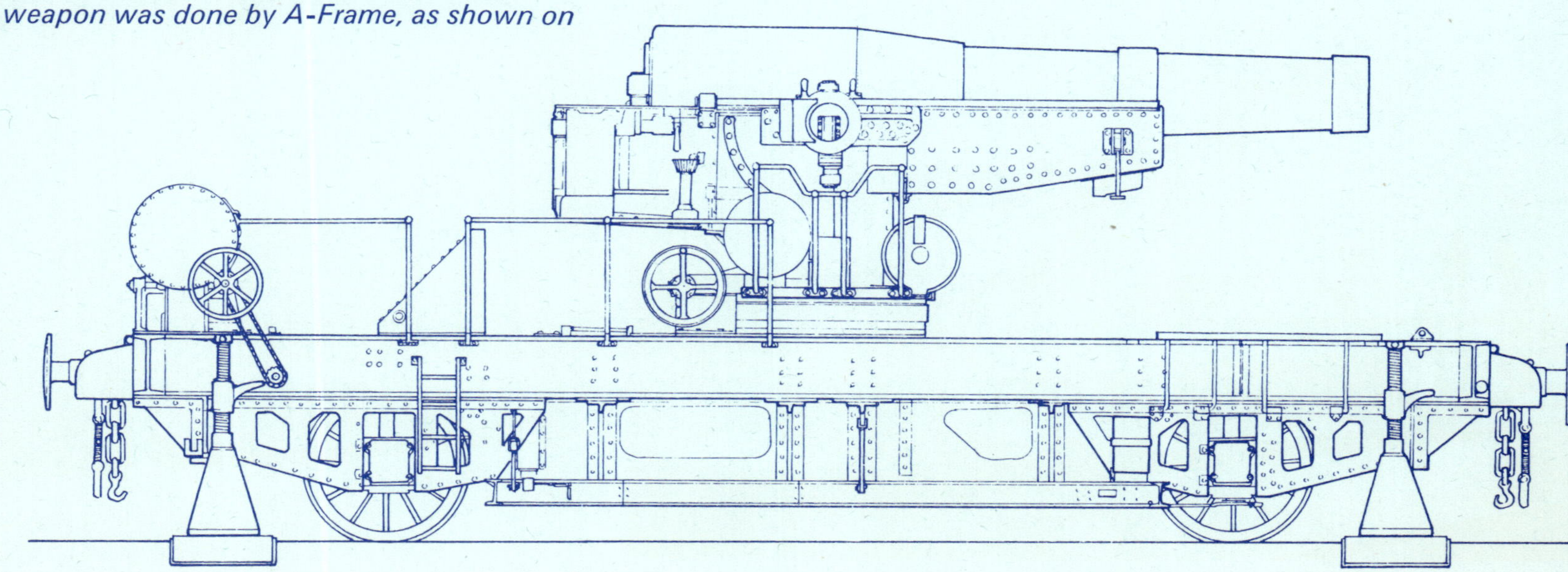

The French 293mm howitzer; (1) in the traveling mode, with the gun and cradle on its transport wagon and the gun mounting with A-Frame behind. (2) A closer view of the gun mounting vehicle. (3) The gun and cradle lowered onto the mounting by using the A-Frame to swing the assembly across.

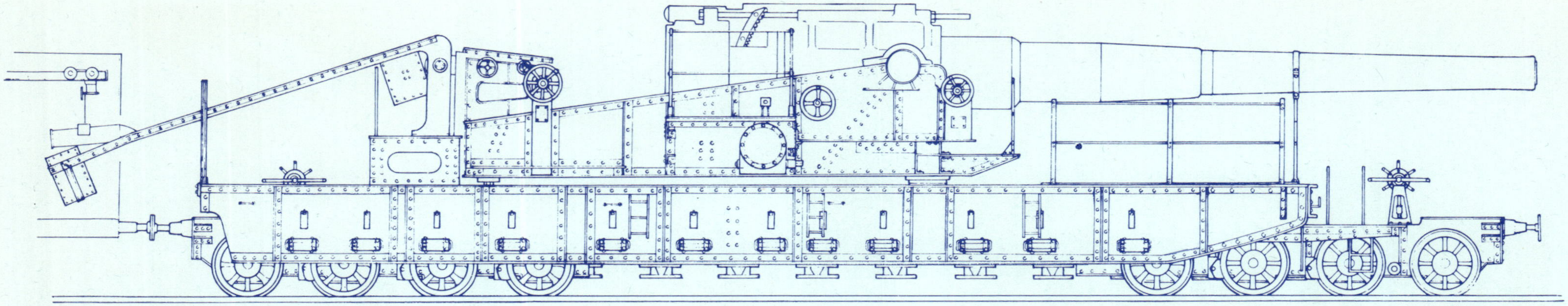

30
305mm Mle 93–06 on Batignolles mounting. This drawing shows the vertical ground spades incorporated in the Batignolles design of ground platform. The weapon also has an unusual and novel design of elevator for conveying the ammunition from the magazine truck to the gun mounting.

A French 400mm howitzer Mle 15 emplaced in a pit for protection against enemy fire. Question: if it takes one man one hour to excavate 25 cubic feet, how long . . . .?

French Canon de 285 Mle 1917. This is in the traveling mode, with jacks and supports up, and appears to be in the process of being trundled back to Germany by its captors using handspikes under the wheels.

31
Materiel de 305 Mle 93–06 sur affut-truc à glissement

Weight: 175 tons (178,000 Kg)
Barrel length: 503 in (12.775m)
Shell weight: 766.5 lbs (348 Kg)
Length: 89′ 0″ (27.2m)
Elevation: 40° Traverse Nil
Velocity: 2610 f/sec (795 m/sec)
Maximum range: 29,965 yds (27,500m)

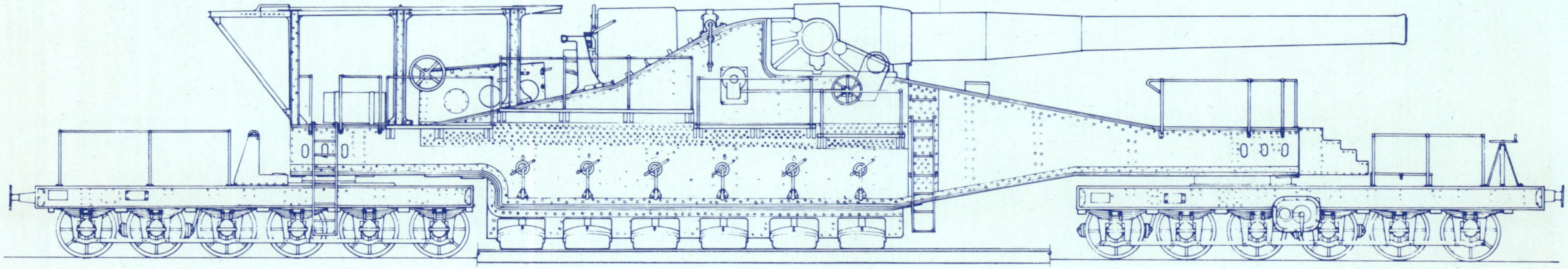

French 340mm Mle 1912 on sliding mount. This well illustrates the track-side reinforcing beams needed with this type of mounting.

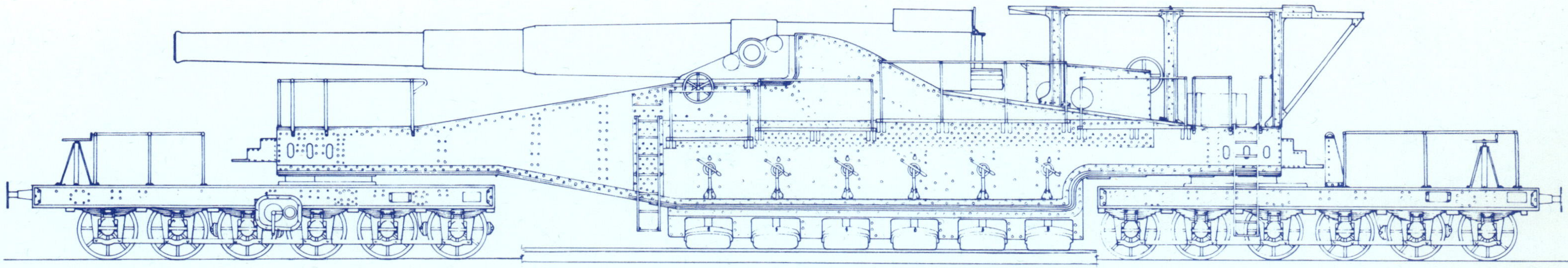

32
Canon de 305 Mle 1906 sur affut-truc à glissement

Weight: 175 tons (178,000 Kg)
Barrel length: 465.35 in (11.820m)
Shell weight: 854.6 lbs (388 Kg)
Length: 89' 2.86" (27.28m)
Elevation: 40° Traverse: Nil
Velocity: 2210 f/sec (674 m/sec)

Maximum range: 30,840 yds (28,200m)

◁ ▷
A French 320mm on sliding mount, new and shiny from the makers. The second photograph shows the breech mechanism; on operating the crank, the skew gears rotate the worm; this rotates the breech block until the stop arm (top) hits a stop block on the breech. Continued working of the crank then withdraws the block to the 'translating tray' until it is completely clear of the breech recess, when further rotation will swing the assembly open until it latches to the gun body. This is a 'continuous motion' breech.

▽ ◁
The French 320mm Mle 1893 ready to fire; the sliding mount has been jacked down on to the firing platform, the breech is open, and a cartridge container is being handled onto the mounting from the magazine truck.

33
Matériel de 320 Mle 1874 sur affut-truc à glissement

Weight: 159.3 tons (162,000 Kg)
Barrel length: 411.1 in (10.442m)
Shell weight: 854.6 lbs (388 Kg)
Length: 85' 0" (25.9m)
Elevation: 40° Traverse: Nil
Velocity: 2212 f/sec (675 m/sec)

Maximum range: 27,120 yds (24,800m)

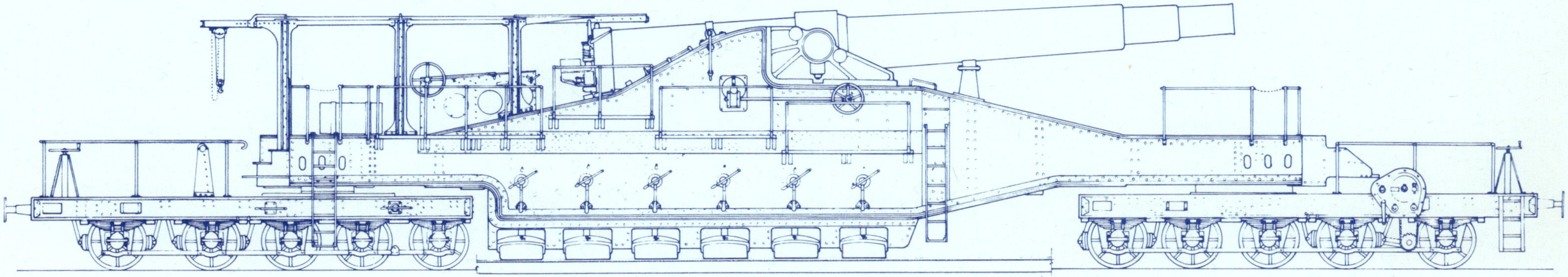

34
Matériel de 400 Mle 15 ou 16 aur affut-truc à berceau
Weight: 137.6 tons (140,000 Kg) *Length: 63' 9" (19.39m)*
Barrel length: 419.3 in (10.650m) *Elevation: 65° Traverse: 12°*
Shell weight: 1412 lbs (641 Kg) or 1982 lbs (900 Kg)
Muzzle velocity: 1740 f/sec (530 m/sec) or 1525 f/sec (465 m/sec)
Maximum range: 17,500 yds (16,000m) or 16,400 yds (15,000m)

35
Obusier de 520 sur affut-truc à glissement Mle 16
Weight: 255.6 tons (260,000 Kg) *Length: 99' 9" (30.38m)*
Barrel length: 328.75 in (8.350m) *Elevation: 60° Traverse: Nil*
Shell weight: 3643 lbs (1654 Kg)) *Velocity: 1477 f/sec (450 m/sec)*
Maximum range: 15,965 yds (14,600m)

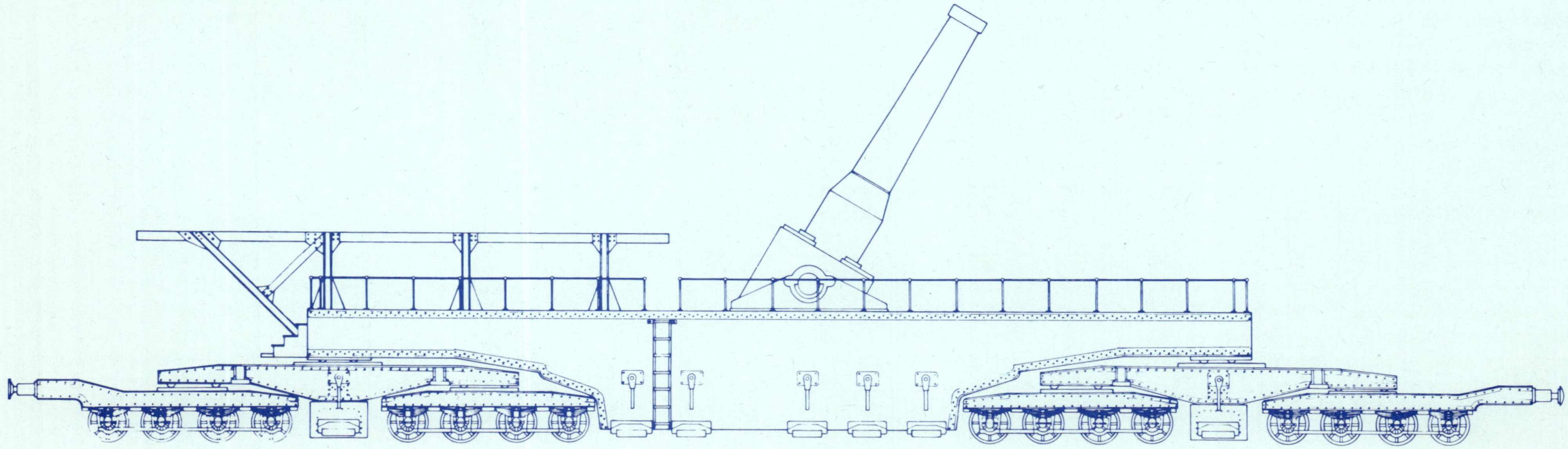

to relieve the springs when firing, and a stout wire hawser led to a ground anchor laid in front of the mounting and prevented any rearward movement of the truck. The gun barrel was mounted in a top carriage riding on an inclined slide bolted to the wagon floor. A hydraulic brake damped down recoil and gravity allowed the carriage to run back to the firing position when recoil ended. This seemed to work in a reasonably serviceable manner—it was not a very powerful weapon, and the mounting took the strain—so a number of antiques were removed and similarly mounted, among them the 24cm howitzer Mle 1876, the 24cm guns Mle 1884 and Mle 1903, the 27.4cm howitzer Mle 1870/81 and the 27.4cm gun Mle 1893. The railway trucks used for these were a heterogeneous collection of two-, three- and five-axle models, the latter being specially built from existing rolling stock components in order to take the weight of the larger weapons.

The 24cm gun was available in some quantity, having been a preferred French caliber for several years—comparable to the British 9.2-inch coast guns—and the next application of the Mle 1884 gun was a much more efficient equipment. The standard coast mounting was lifted out of its concrete and dropped bodily on to a racer ring set into a reinforced well base flat wagon to make a platform mounting with 360 degrees traverse. Two small ready-use magazines were placed at each end of the wagon in a similar arrangement to that of the 164mm guns. The gun mounting was of the 'Vavasseur' pattern, in which the top carriage was formed into two inclined planes at its top, up which the gun recoiled in a small subcarriage, damped by hydraulic buffers and returned to the firing position by a combination of gravity and spring power; it was simply a cleaned-up version of the old sliding mounting used with garrison guns, but the improved buffing arrangements kept the maximum recoil stroke down to about 1.5 metres.

As usual, successive models of the 24cm gun became heavier and more powerful and the 24cm Mle 93/96 demanded a somewhat more robust mounting to deal with the greater weight and energy, but apart from an increase in size it was essentially the same basic design as the earlier type, a 360-degree traverse platform on a well base wagon. This weapon saw one of the first applications of the prepared platform anchored mounting which was later improved into the Batignolles design.

The 27.4cm guns Models 1887, 1893 and 1917 were the smallest calibers adapted by Schneider to their pattern of non-recoil sliding mount, and they can be taken as being representative of the class. The basic structure consisted of two heavy side plates rigidly connected by an elaborate system of cross-transoms to make a rigid box. The side members were shaped so as to accept the gun trunnions in a reinforced area, with a solid steel bushing to take the recoil blow, and the basic box was supported at each end by a suitable bogie. The necessary arrangement of cross-beams and screw jacks was fitted to the underside of the center section and platforms as required for the service of the gun were provided. In addition to the 27.4cm guns, 28.5cm guns, 30.5cm guns and howitzers, 32cm guns, 34cm guns, 37cm guns and howitzers were similarly mounted, the only significant change being the size of the bogies necessary to carry the extra weight. Five or six axles were usual, but some of the 34cm and 37cm weapons used two four-axle bogies at each end, tied together by a span bolster to which the mounting was pivoted. Such an arrangement, as well as distributing the weight, gave better articulation to the mounting and allowed it to be maneuvered on the somewhat sinuous tracks laid behind the lines. It may be noted here that the European standard loading gauge imposed a weight limit of 17,000 kilograms per axle, which accounts for the varying wheel arrangements by simple arithmetic.

The 34cm gun Mle 1912, after being installed in a sliding mounting, was also placed in a cradle mounting; two very similar models of this gun existed, one with a rifling twist of four degrees (one turn of the shell in 45 calibers) and the other with a twist of six degrees (one turn in 30 calibers). This may sound a trifling difference, but with a slight alteration in propelling charge the power of the six-degree rifled model, expressed as the muzzle energy, was increased from 18,900 metric tons to 20,550 metric tons and the range increased by rather more than four miles. This additional power meant a heavier recoil blow and the sliding mount was adjudged insufficient to withstand it, so a cradle mounting had to be made. Basically it was the Schneider sliding mount but with the gun carried in a ring cradle trunnioned to the mounting side plates. Four hydraulic recoil cylinders and one hydro-pneumatic recuperator cylinder were built on to the cradle. The latter cylinder was operating at 2300 pounds per square inch, an exceptionally high pressure indicative of hurried design.

With the perfection of the Batignolles mounting in 1917, the 30.5cm gun Mle 93/06 and the 37cm howitzer Mle 1887 were constructed to this pattern and became more or less the preferred types, replacing the earlier Schneider sliding mountings as they wore out—although many of these survived to the postwar years.

In order to produce a really heavy weapon, the St Chamond company took a worn-out 34cm naval gun Mle 1887, shortened it and rebored it to 40cm caliber to make a howitzer. The mounting was of their own design and was a simple flat-topped girder bearer mounted on bogies and carrying the gun in a cradle and top carriage. This unit was pivoted at the rear to allow limited top traverse. The mode of use was based on the Batignolles system, the mounting being lowered to the ground before firing, though the majority of the recoil force was absorbed by the hydro-pneumatic recoil system. It was a very clean and efficient design, but the weapon was insufficiently powerful for its size and few appear to have been made.

The largest French rail gun design was the Schneider 52cm howitzer. As already mentioned, this was a political necessity rather than a military requirement. Two were built, but they saw little use, for dragging 255 tons into place and spending three hours emplacing it in order to fire nine miles was hardly cost-effective. It is notable that this was the only French railway weapon which was actually designed during the war instead of the barrel being adapted from prewar reserve stocks.

France ended the First World War with an impressive collection of railway artillery, probably as much as the remaining combatants put together, but as must be obvious from the foregoing, much of it was only fit for the welder's torch once the war was over, and that was where it went. However, considerable stocks of the Batignolles weapons were retained, together with the various cradle mounted guns, the 164mm platform weapons and a number of the sliding mount designs. In the postwar years one new design was developed, a 62-caliber long model of the 34cm gun on a Batignolles mounting with cradle, but very little is known of it. It is believed that no more than the prototype was ever built. When the Second World War began, most of these weapons were brought into prepared positions to augment the Maginot Line defenses, from which they occasionally fired interdiction and harassing missions. In 1940 they were captured by the Germans. Most were scrapped out of hand, but a number of the more efficient designs were taken over by the occupation forces and used to stiffen up the coastal anti-invasion defenses.

British Railway Guns

If we except the mounting of 4.7-inch guns on railway trucks at Alexandria in 1882 and as a part of the column for the Relief of Ladysmith in 1900, both of which were simply field expedients for the rapid movement of weapons, the first official British recognition of the use of guns on railways appeared in November 1900. Probably as a result of the Boer War the Director General of Ordnance raised the question of producing a mounting to enable the six-inch 30-cwt howitzer to be moved on a 2′ 6″ gauge siege railway and fired from the track. By March 1901 a design had been produced which mounted the howitzer in an ex-coast gun central pivot mounting with cradle recoil, allowing an arc of traverse of 180 degrees, the gun firing over one end of the truck only. The other end mounted a small magazine to carry 16 rounds of ammunition. The design of mounting was highly ingenious. The wheel axles were mounted on eccentric blocks coupled to a screw gear so that revolving the screw would turn the eccentrics to lower the mounting until it rested on baulks of timber placed between and alongside the track. Further movement of the eccentrics then lifted the wheels completely clear.

An alternative design with shorter recoil and with the magazine removed so that the gun could fire over an arc at each end of the wagon was also drawn up so that the powers-that-be could make a choice. Unfortunately, it wasn't to be as simple as that. The Director-General of the Ordnance turned down the designs on the score of their failing to give all-round traverse; the Superintendent of the Royal Carriage Department submitted fresh designs; the Inspector of Guns and Fortifications now appeared and objected that the wheelbase was too great for a rigid frame and that bogies should be used, and that the eccentric system should be changed to using screw-jacks at each corner. He also demanded better couplings, brakes, and spring suspension. The Superintendent said well, all right, but I only intended it for slow movements, not for express trains. After this it got even more involved with such odd appointees as the Inspector of Iron Structures making comments and raising technical objections. Eventually, late in 1902, the Inspector of Guns and Fortifications pointed out that the whole thing had mushroomed to a complication far exceeding the DGO's original idea, and what was wrong with simply carting the standard field howitzer around on a flat wagon and trundling it off to the trackside when necessary, 'for which he understood satisfactory appliances now existed'.

This douche of cold water had its effect; 'the Director-General of Ordnance stated that the question might be dropped'. And that was that on the rail gun front for another thirteen years or so.

When the war broke out in 1914 Britain was not quite so badly off for heavy artillery as the French, and there was no rush to develop railway weapons. It was not until the middle of 1915 that the demand arose, and it was solved in much the same way as the French had done it—utilizing old weapons held in reserve for coast defense or naval armament. It was customary in those days to manufacture an excess of gun barrels over requirements and place them in store against the day when the service weapons wore out, and it was this hoard which was first tapped. In order to mount them the usual British system was adopted; to call on the specialist gunmakers, give them the problem and let them get on with it, a system which generally produces the best guns. The Elswick Ordnance Company were approached with a request that they mount some 9.2-inch

36 ▷
9.2-inch Gun Marks 3 to 6 on Mounting Railway Truck Mark 1
Weight: 60 tons (60,960 Kg) (Approx.) *Length: 40′.029″ (12.196m)*
Barrel length: 310.0 in (7.874m) *Elevation: 28° Traverse: 10°*
Shell weight: 380 lbs (172.4 Kg) *Velocity: 2100 f/sec (640 m/sec)*
Maximum range: 12,900 yds (11,795m)
(The exact weight varied between 54 and 60 tons, depending upon which mark of gun was fitted to the mounting.)

The first British 9.2-inch Railway gun, built by the Elswick Ordnance Company in 1915.

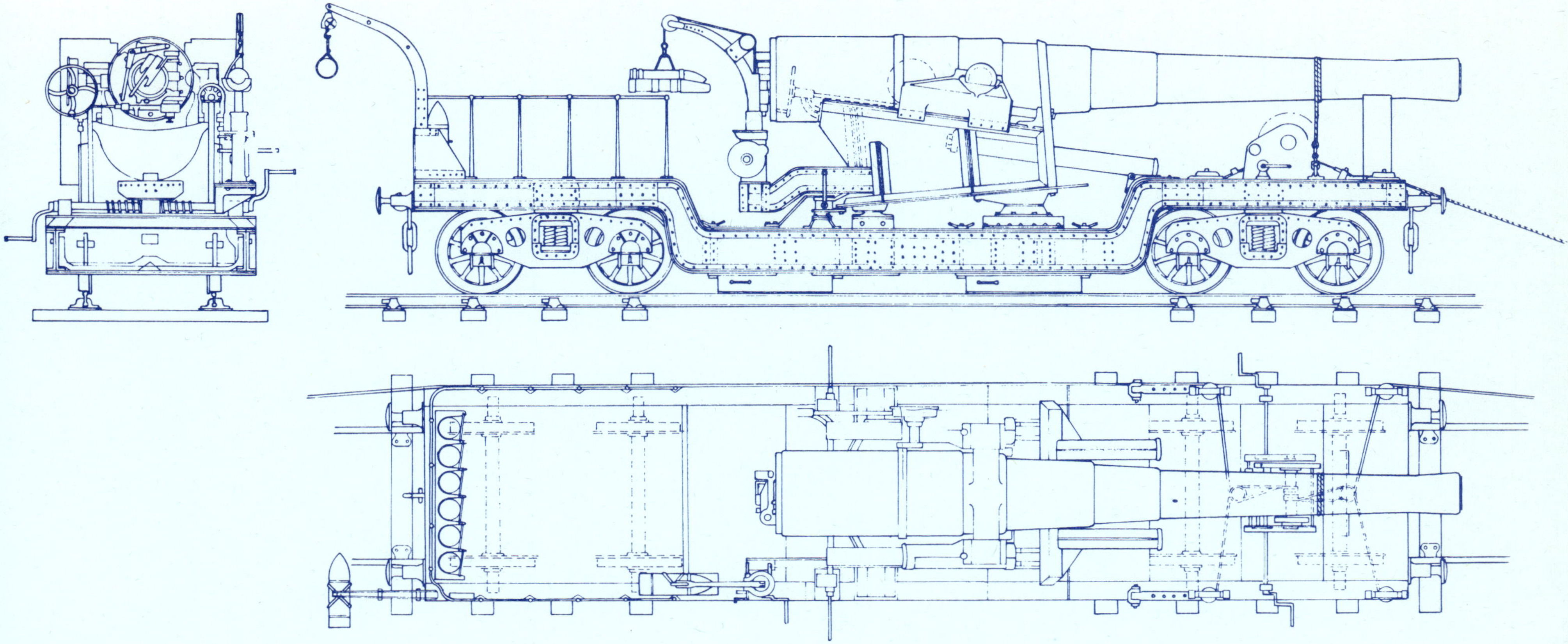

37 ▽
9.2-inch Gun Marks 3 to 6 on Mounting Railway Truck Mark 1A. This was virtually the same as the previous model but with a lengthened truck body.

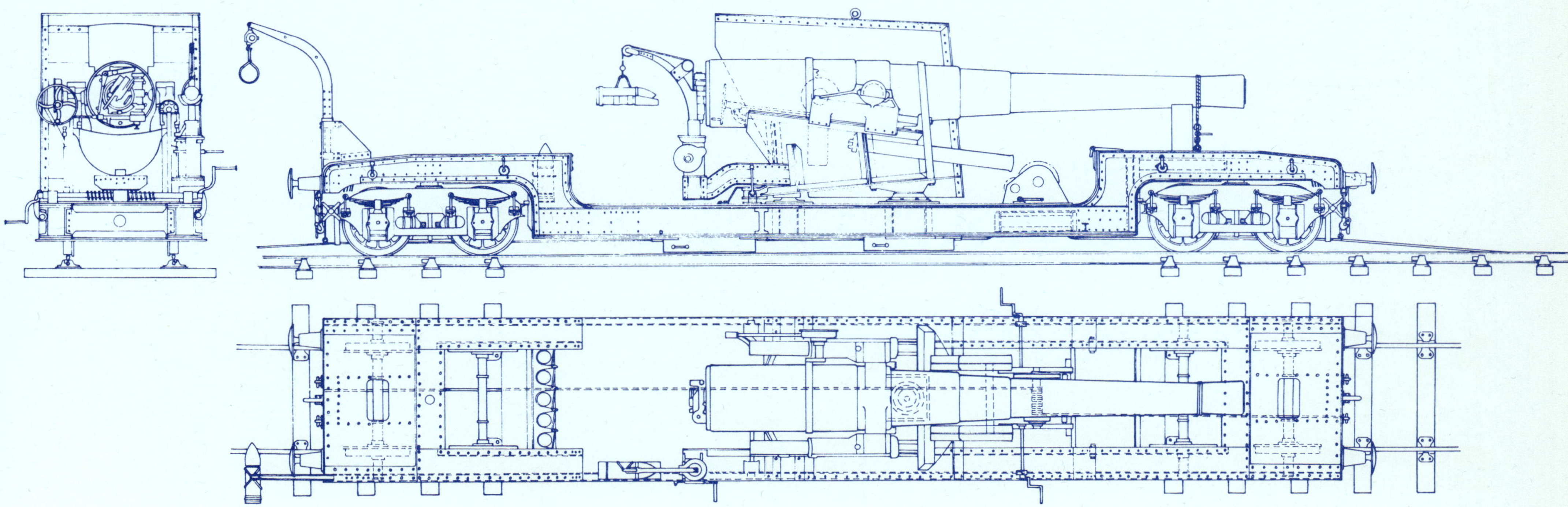

guns, and they soon produced a design using a well-base double-bogie truck carrying a Vavasseur slide mounting. This type of mounting was readily available, since many naval broadside guns and coast guns had used them, so that a suitable design was rapidly put together from standard components. The truck was little more than a normal commercial well wagon, suitably strengthened, and the mounting as a whole was of the rolling pattern.

The equipment was reasonably serviceable, but suffered from lack of range due to the design of the Vavasseur slide mounting which restricted elevation to 28 degrees. Early in 1916 the company were asked if they could alter the design so as to reach a higher elevation and thus obtain more range. In a matter of days a fresh design was produced in which the front end of the recoil slide was raised by inserting packing pieces, and this altered the recoil geometry to allow 35 degrees of elevation. The necessary components for the modification were made up into kits of parts and sent to France where the mountings in service were rebuilt to the new pattern on the spot.

After producing these mountings the Elswick Company knew exactly what to do next. After many years of experience in supplying guns to the British Army and Navy, they knew very well that in a short time the customers would be back for something more powerful, lighter, better engineered, and all the other familiar demands made by soldiers to gunmakers since the Battle of Crécy. They then set about designing a fresh pattern of 9.2-inch gun mounting from the rails up, which was produced in June 1916, a double-bogie straight-back truck with a Vavasseur slide of improved pattern mounted on a racer to give all-round fire.

Outriggers were attached to the mounting, with heavy spades at the outer ends, and the truck was secured to the track with rail clamps for firing. A screw mechanism in the bolsters could be used to lift the bogies and remove the weight and firing stresses from the suspension. The guns which eventually found their way on to this mounting were an odd collection. The 9.2-inch Mark 10 was the 'approved pattern', this being the most modern and powerful design available, but since numbers in reserve were earmarked for the fleet, not all the mountings could be supplied with them. The Elswick Company had two 9.2-inch coast guns which they had built to the specification of the Australian Government, and these were appropriated under the terminology 'Mark 10T'. The Vickers Company had four guns which they had built for sale abroad in prewar days, and these too were adapted to the railway mount, though since they were much different to any existing service pattern they

38
9.2-inch Gun Marks 3 to 6 on Railway Truck Mark 1A Modified. The front end of the slide was packed up to allow the gun to reach 35 degrees elevation, thus increasing the maximum range to 16,000 yards.

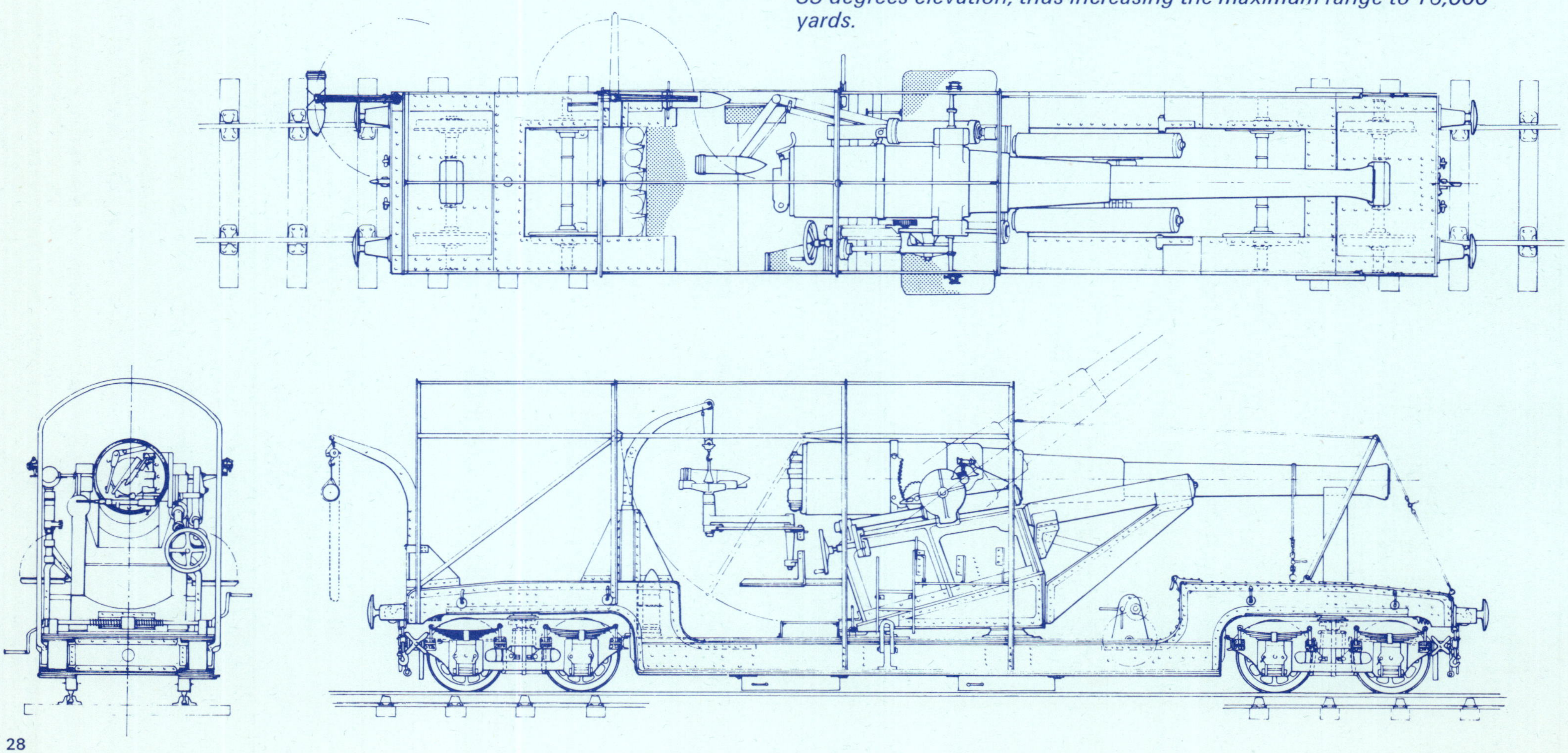

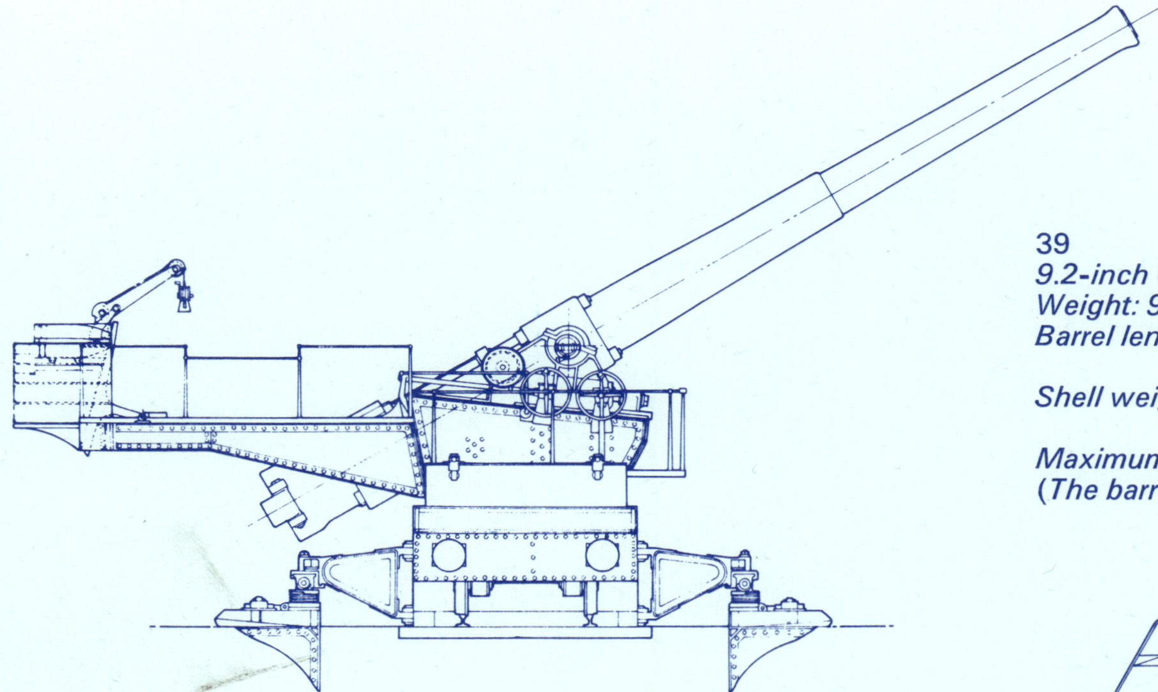

39

9.2-inch Gun Marks 10, 10T or 14 on Railway Truck Mounting Mark 3

Weight: 90.6 tons (92,056 Kg) Length: 46' 10" (14.27m)

Barrel length: (Mks 10, 10T) 442.35 in (11.226m)

Elevation: 30° Traverse: 360°

Shell weight: 380 lbs (172.4 Kg) Velocity: 2700 f/sec (823 m/sec)

Maximum range: 21,000 yards (19,200m)

(The barrel of the Mark 14 gun was 427.025 inches (10.845m) long).

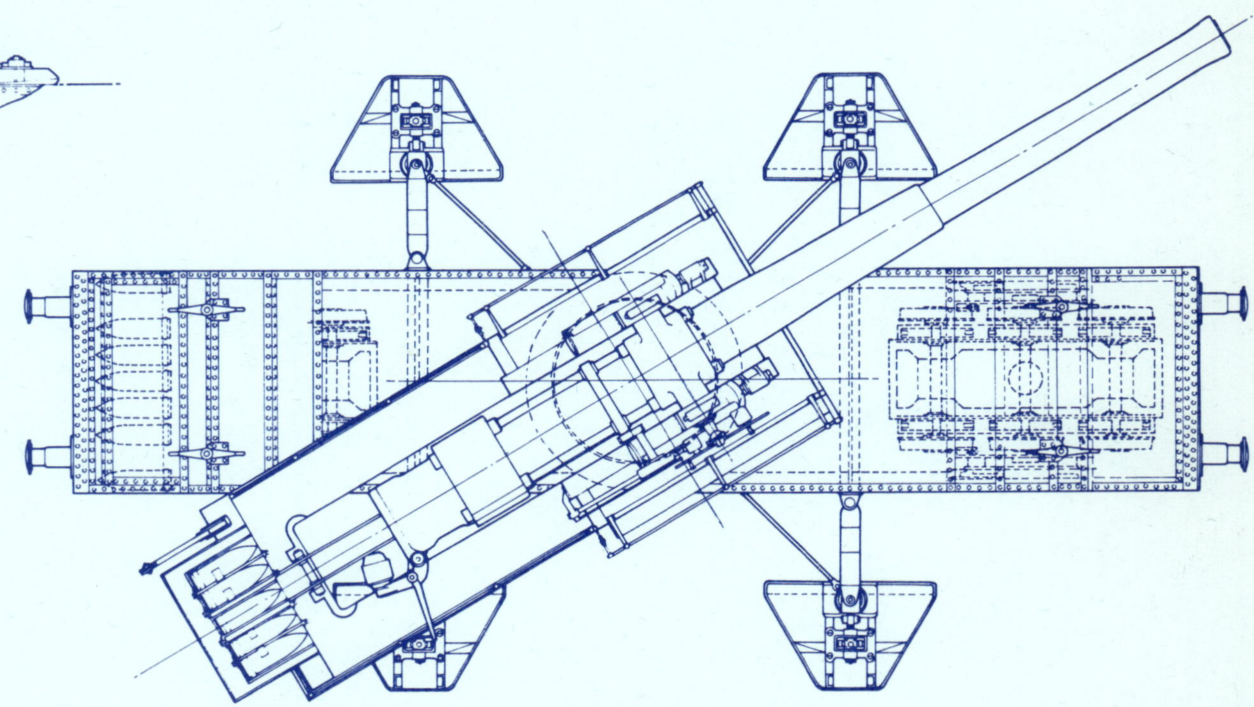

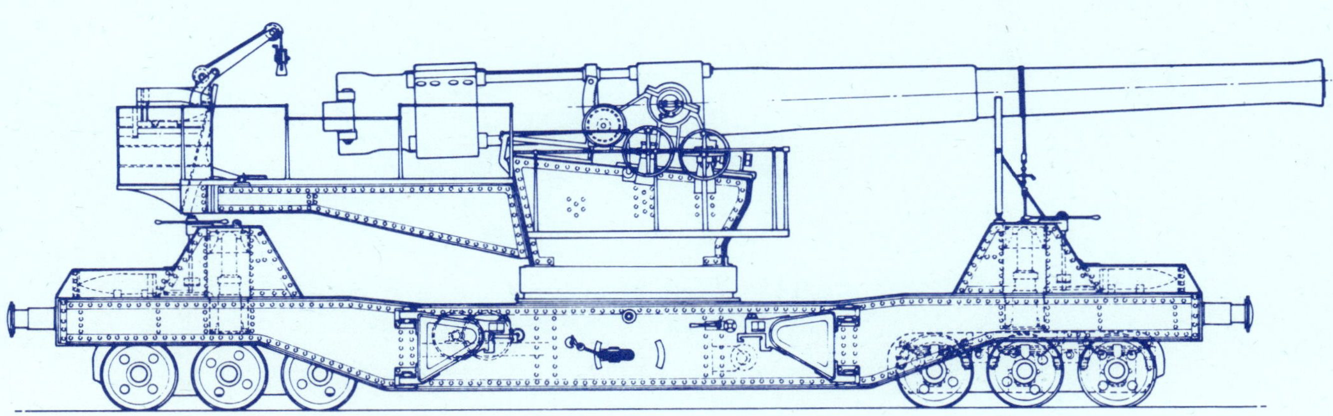

became the 'Mark 14' gun.

Although these designs were a great improvement on the earlier mounting, there was still a range restriction. Due to the design of the guns it was necessary to mount the trunnions well forward on the chase to balance the muzzle and breech weights and thus balance the gun, so that it was capable of being elevated by hand power. Since there was a likelihood of the breech fouling the mounting, elevation was restricted to 30 or 35 degrees, depending on which mark of barrel was mounted. A design was drawn up with the gun in a ring cradle with buffer and recuperator cylinders above and below the cradle to give a short axial recoil, but this design was not taken up, probably because cutting the recoil so short would have upset the stability of the mounting. The solution finally adopted was to design a new gun barrel to fit the mounting. This was somewhat shorter than the original Mark 10 pattern and was made with the breech end much heavier so that the trunnions could be moved back and thus allow the gun to reach 40 degrees of elevation and add another mile to the range. This gun entered service as the Mark 13 and remained the standard railway gun for many years.

An interesting proposal was put forward in 1918 which involved taking an elderly ten-inch naval gun and re-lining it to 9.2 inches. The guns were originally built by the Elswick Company for arming a warship ordered by the Chilean Navy, but due to internal dissension the Chilean Government failed to take delivery and the ship was taken into the Royal Navy as HMS *Swiftsure* in 1904. A complete outfit of reserve guns had been made and these were now considered surplus to requirements, since the navy had no intention of retaining *Swiftsure* after the war (it was scrapped in 1920). A design of '9.2/10-inch Mark 6' gun was approved, which would have been an exceptionally powerful weapon. But it was almost twice the weight of the existing 9.2-inch guns, and for this reason it was decided that it would have to be mounted into the 12-inch gun mounting. Although the design was approved the war ended before any work could begin, and the project was abandoned, the design being formally declared obsolete in 1920.

The next Elswick design once the original 9.2-inch guns were out of the way was the production of a heavier howitzer. The standard heavy howitzer of the British Army was the 9.2-inch, while the Royal Navy owned a few 15-inch manned by Royal Marines in use on the Western Front. This latter weapon had been a private venture and the Army

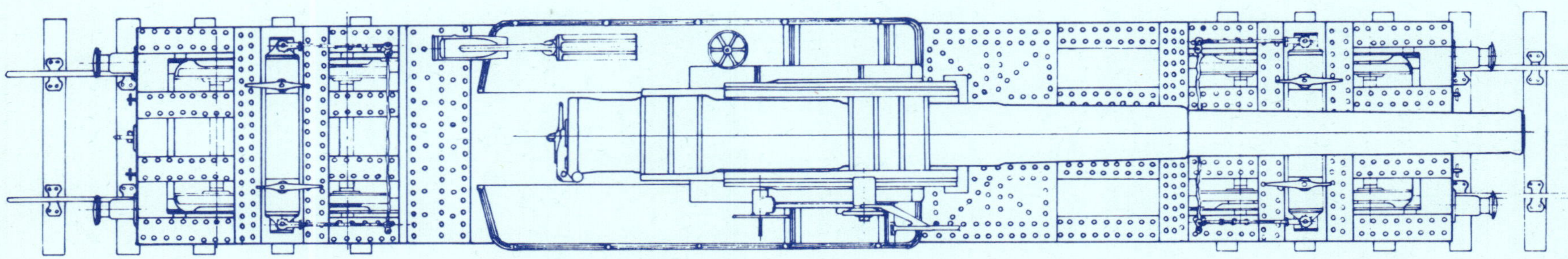

40
9.2-inch Mark 10 gun as designed with an axial recoil system in a ring cradle. The design did not enter service. Note that the end view has the standard British loading gauge superimposed.

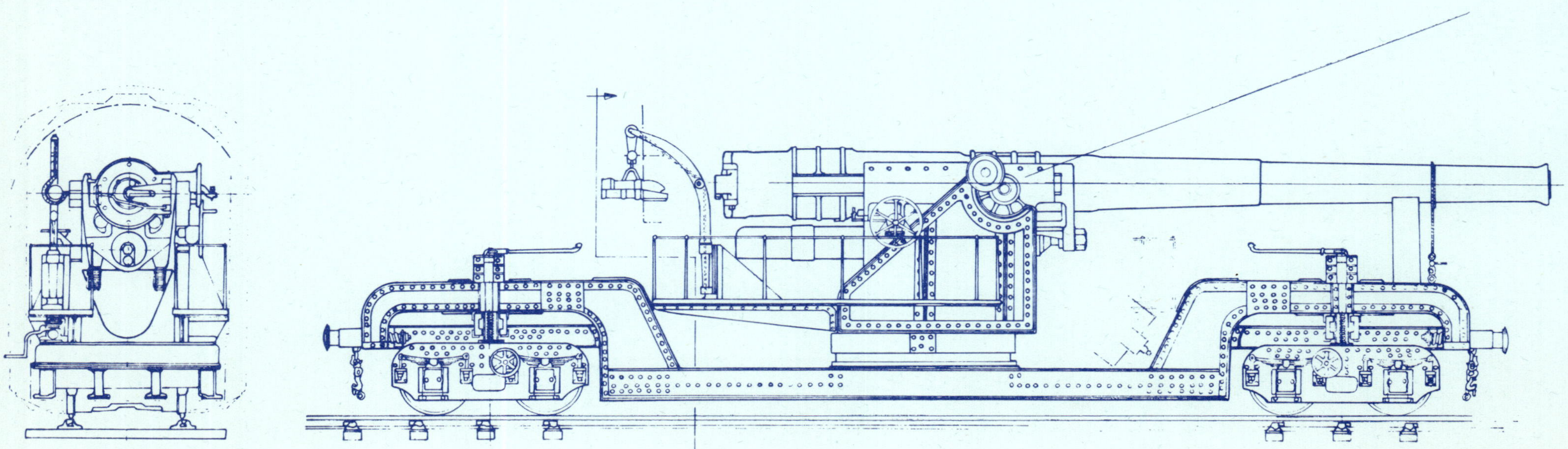

preferred not to know about it, while the 9.2-inch was not considered worth the trouble of mounting on rails. Splitting the difference gave twelve inches as a nice round figure, and the Elswick Company were asked to put such a weapon on rails, while Vickers were given the job of producing a road-mobile version.

Elswick did the sensible thing in the circumstances—they took a look at the existing 9.2-inch howitzer built by the Coventry Ordnance Company and then more or less scaled it up to the required caliber. Details of the recoil system and breech mechanism differed of course, since Elswick had their own ideas on such subjects. The mounting was a well base truck with a racer ring so that the howitzer could be traversed through 360 degrees, but to avoid complicating the issue with outriggers, it was restricted to firing within a 40-degree arc over each end of the truck. In action the mounting was lowered by screw mechanisms in the bogie bolsters until the base of the mounting rested on baulks of timber laid between the rails and the bogies were relieved of weight. Rail clamps were then applied to hold the assembly steady during firing.

While this was not a bad beginning, Elswick knew perfectly well that before long an improvement would be wanted. They immediately set about designing a new barrel which would allow a heavier charge to be fired to reach a greater range. The mounting was little changed except to strengthen it to resist the greater stress, and the complete equipment, known as the Mark 3 howitzer, was ready in late 1916 and gradually superseded the Mark 1 design in service.

But the customers were still not satisfied; complication or not, they demanded the ability to fire at any angle to the track. The Elswick men picked up their pencils once more and produced the Mark 5 model. (The Marks 2 and 4 were Vickers designs of road-mobile weapons.)

41
The first design for mounting the Mark 13 pattern 9.2-inch gun. This was not considered sufficiently stable and the loading platform was inconvenient.

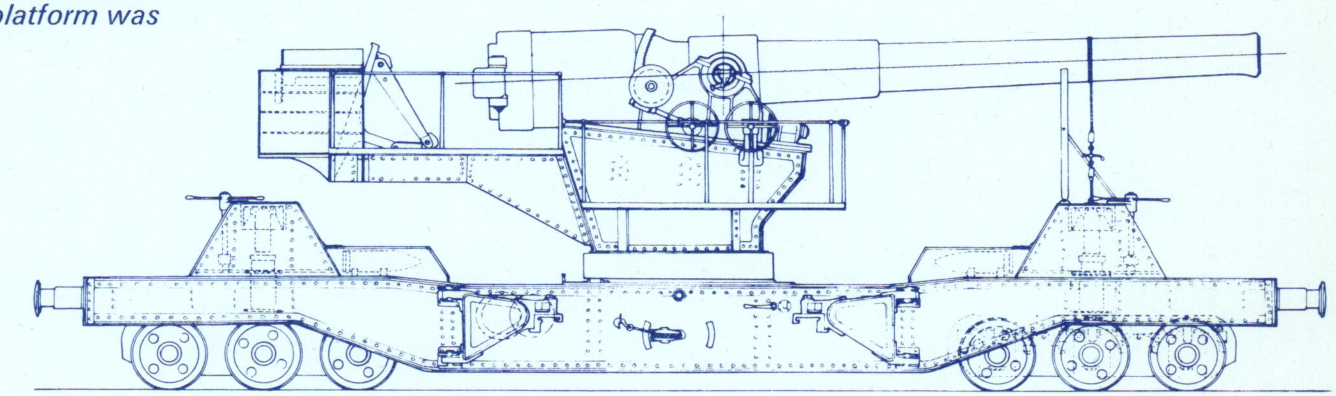

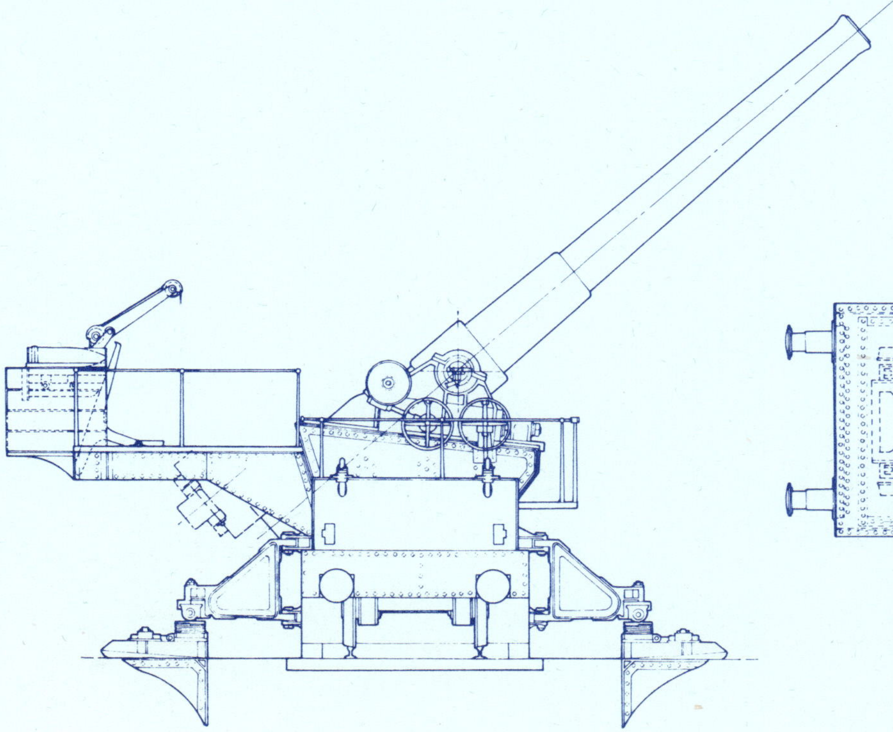

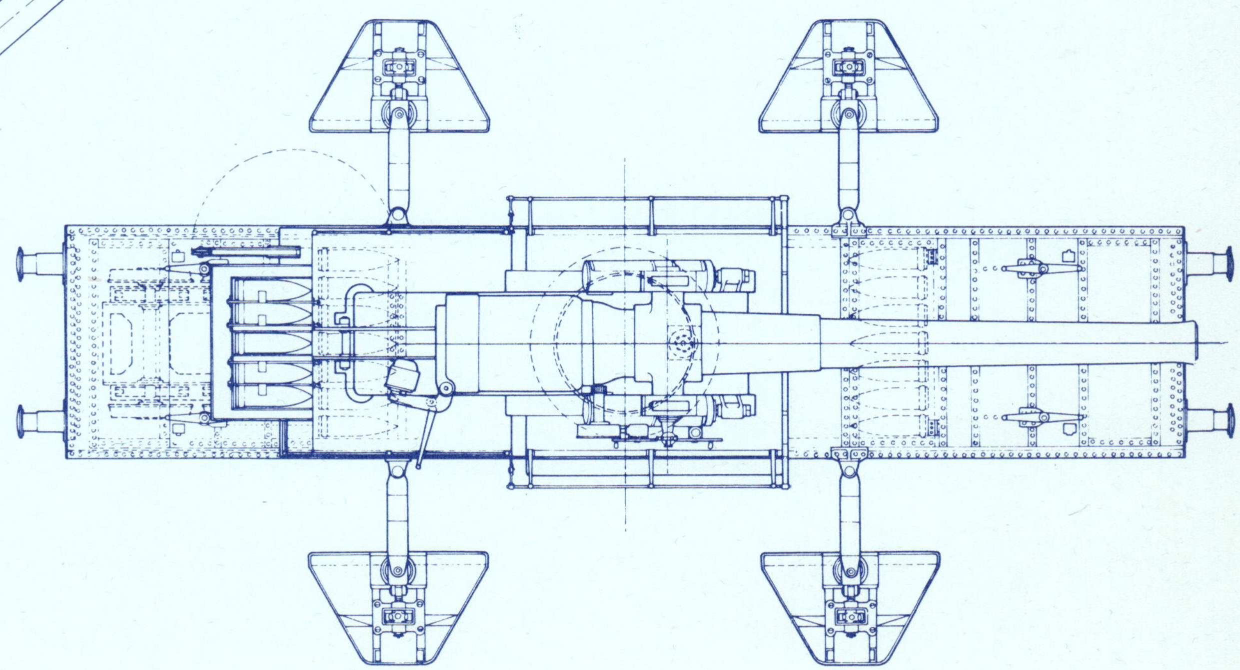

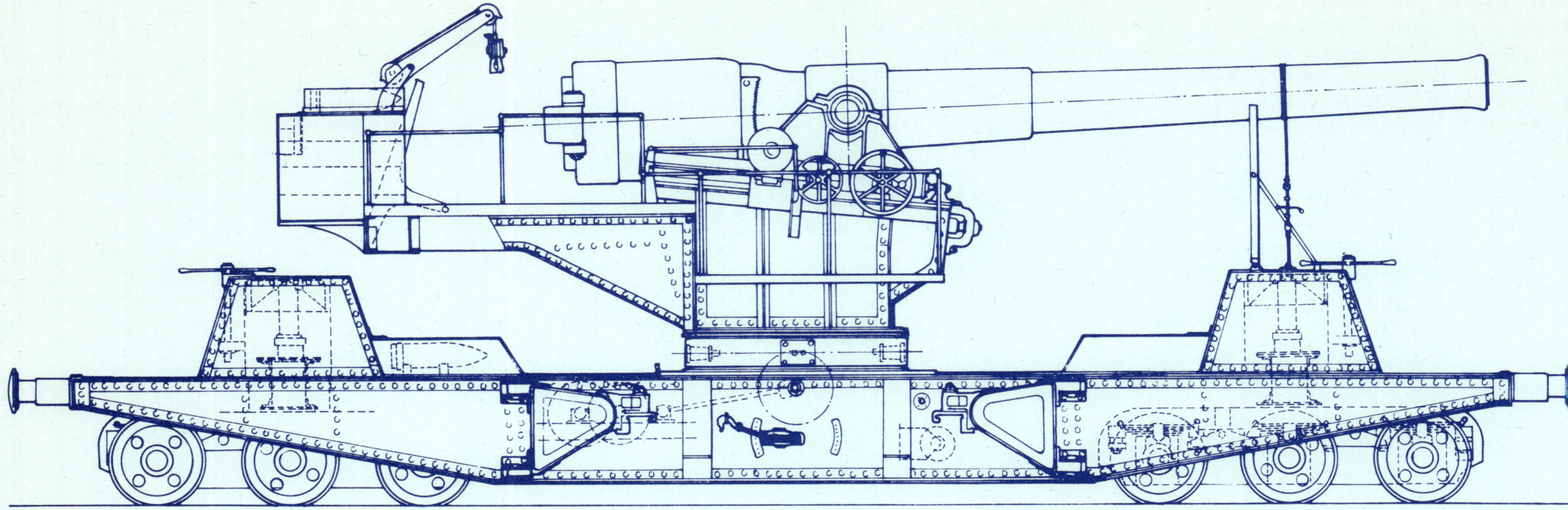

42
The second design of Mark 13 9.2-inch gun mounting produced by the Elswick Company. This was approved, subject to small changes principally to suit manufacturing convenience.

43
An emergency design, using the 9.2-inch Mark 13 gun fitted to the original Mark 1A truck mounting. This was approved for production should there be undue delay in the provision of the final version, but none except the prototype were ever built and the design never officially entered service.

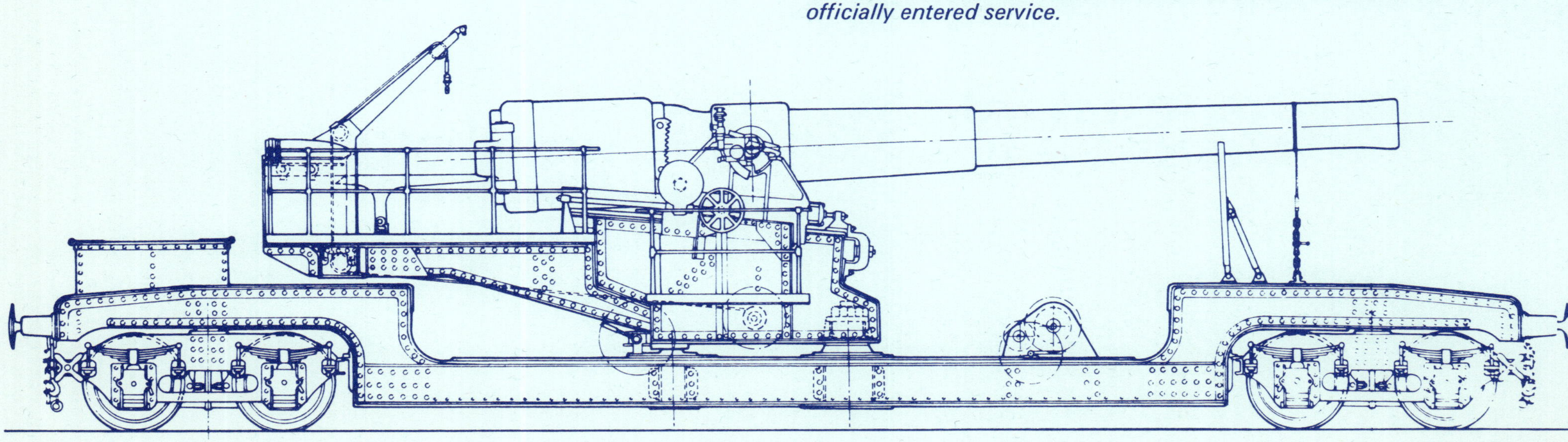

The howitzer was more or less the same as the Mark 3 but the cradle and recoil system were redesigned to lower the silhouette and the center of gravity. The mounting was a turntable on the same well base truck, but with a loading platform carried on the turntable so that the gunners had somewhere to stand when the howitzer was firing across the track. Outriggers, spades and rail clamps were the standard method of anchoring the platform when firing, though an interesting field expedient to be seen in photographs taken in France involved the use of cables and ground anchors. Both the Mark 3 and Mark 5 howitzers were very successful designs and they remained in service with the British Army until 1945, although they saw no active use after 1918.

The 12-inch gun designs used in British service form a peculiar tale of private enterprise and inter-service buck-passing. The story opens in December 1915 when a somewhat astonished Director of Artillery informed the Ordnance Board that he had just been presented with a *fait accompli;* the Admiralty had taken two 12-inch guns from the reserve earmarked for HMS *Cornwallis,* given them to Vickers with instructions to make suitable mounts, waited until the weapons were almost complete, and then informed the army that they were about to receive two railway guns for use in France. One might think that the Admiralty were taking a gamble in using reserve fleet guns, but in the event it was justified; *Cornwallis* was torpedoed off Malta in 1917, so the reserve guns were never needed.

Just what possessed the Admiralty to do this is by no means clear. It may be that with their 15-inch howitzer battery, they felt that the army wasn't doing enough in the big gun line and needed a nudge. One is inclined to discern the hand of Winston Churchill in this somewhere.

The Vickers mountings were quite unique in appearance, in that the

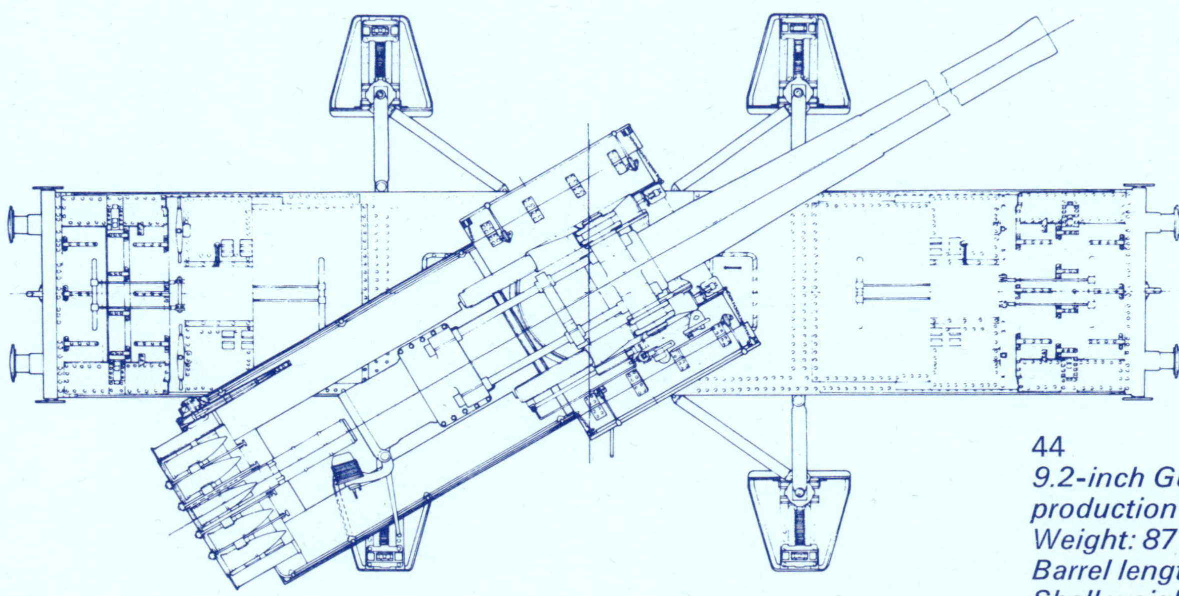

44
9.2-inch Gun Mark 13 on Mounting Railway Truck Mark 4. The final production design which entered service
Weight: 87 tons (88,400 Kg)
Barrel length: 335.025 in (8.509m)
Shell weight: 380 lbs (172.4 Kg)
Length: 47′ 0″ (14.326m)
Elevation: 40° Traverse: 360°
Velocity: 2100 f/sec (640 m/sec)
Maximum range: 22,600 yds (20,665m)

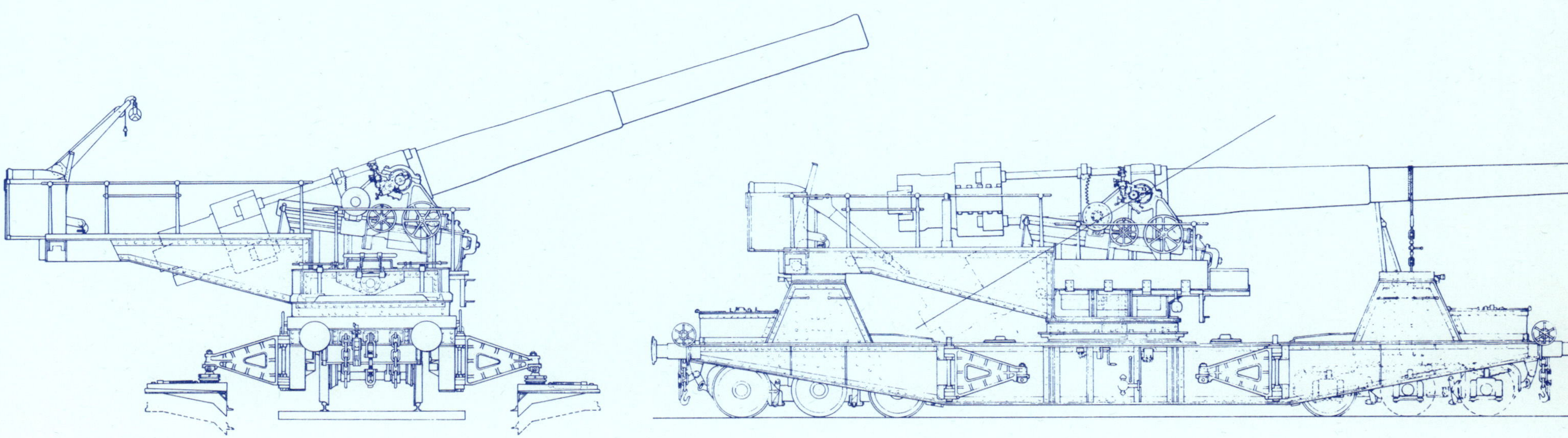

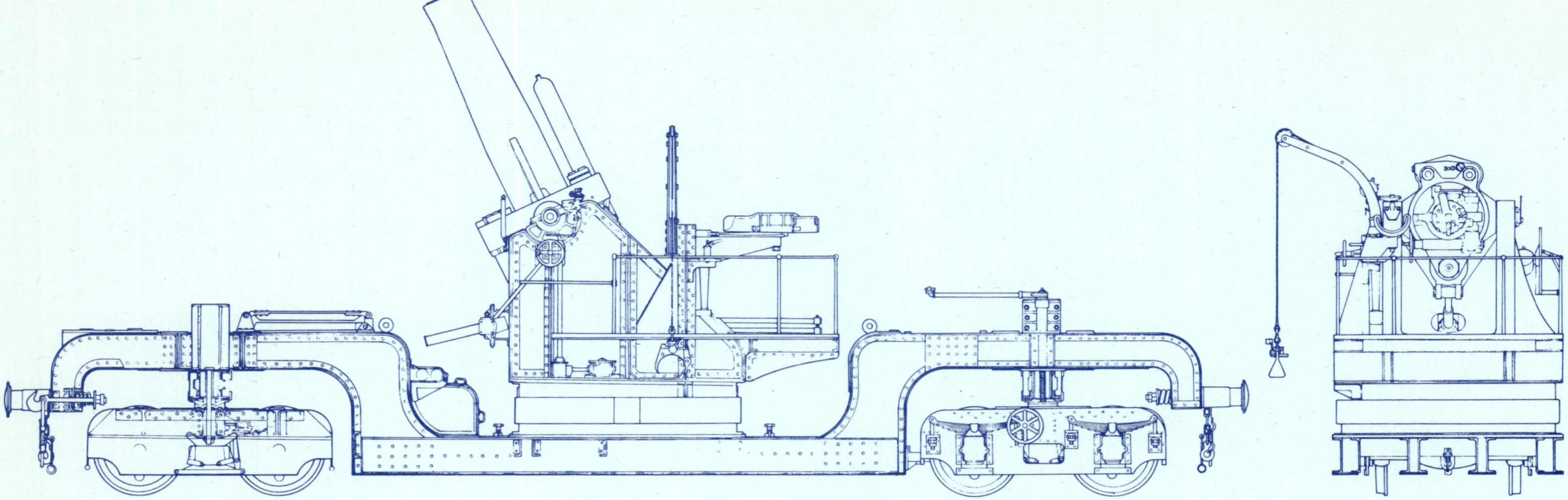

45
12-inch howitzer Mark 1 on Mounting Railway Truck Mark 1
Weight: 57.85 tons (58,826 Kg)
Barrel length: 162.2 in (4.114m)
Shell weight: 750 lbs (340 Kg)
Maximum range: 11,130 yds (10,180m)
Length: 27' 0" (8.2m)
Elevation: 65° Traverse: 360° possible, but only fired within 20° of centre-line
Velocity: 1175 f/sec (358 m/sec)

46
12-inch howitzer Mark 3 on Mounting Railway Truck Mark 2
Weight: 60.75 tons (61,976 Kg)
Barrel length: 225.3 inches (5.715m)
Shell weight: 750 lbs (340 Kg)
Maximum range: 15,000 yards (13,715m)
Length: 40' 0" (12.192m)
Elevation: 65° Traverse: 360° possible, but only fired within 20° of centre-line
Velocity: 1468 f/sec (447 m/sec)

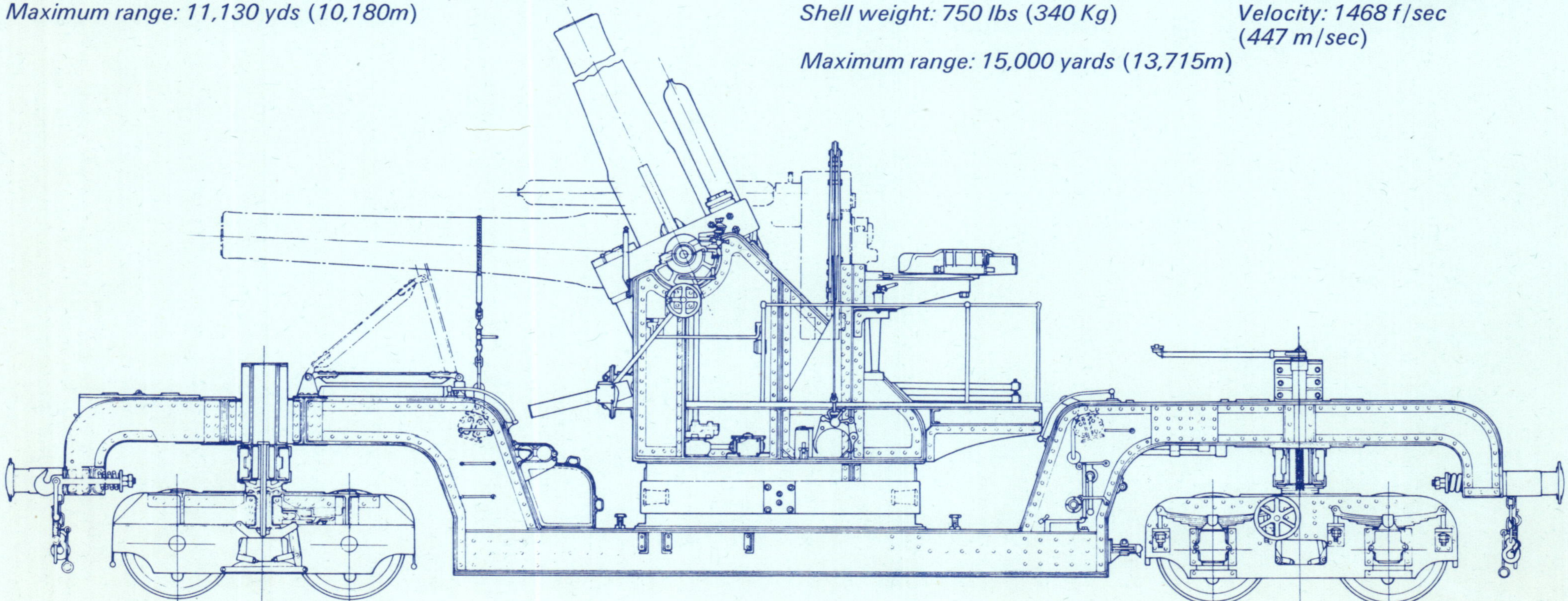

British 12-inch howitzers Mark 3 in action in France. The gunlayer is loading the gun preparatory to firing, while the next shell has just been hoisted inboard.

wheels are carried on inside frame bearings and the whole wheel is visible. The two mountings, while of the same general design, exhibit small variations in details of assembly and construction which indicate that the design was modified as it was built, minor deficiencies in the first mounting being corrected in the second. With the greatest respect for Vickers, who undoubtedly did the best they could in the time available—and remember that at this time nobody else had ever built a rail mounting for such a powerful weapon—these two must be the ugliest railway guns ever built.

No matter. They were built for shooting, not exhibition, and they were duly delivered to the 53rd Siege Battery, Royal Artillery, in late 1915 and went into action forthwith, flinging 850 pound shells to a range of over 16 miles. Their accuracy and effect were such that the War Office decided to have another pair. The Elswick Company were presented with two more barrels and a list of comments by the users in France, and invited to produce their idea of a 12-inch railway gun. The result was a much tidier design, of the usual girder construction carried on conventional outside-framed bogies. These weapons were completed and in action by August of 1916, a commendably quick time, since the original request was only made in February of that year. Both the designs of 12-inch were rolling mountings, and they were retained in service after the war until being declared obsolete in 1930.

A later design of 12-inch gun was mooted in 1918. When the 14-inch gun was in the design stage it was thought that the mounting might well be adapted to carry the most powerful 12-inch gun available, the Mark 11. This gun had been developed to arm HMS *St Vincent* in 1907 and had a range of $17\frac{1}{2}$ miles with an 850 pound shell. Plans were drawn up for the conversion of these guns to suit railway mountings, but the end of the war arrived before the design work was complete. The project was reconsidered shortly after the Armistice, when many of the late war designs were given a critical review. It was felt that if a heavy gun was wanted the 13.5-inch would probably be better value; the 12-inch Mark 11 design was thereupon abandoned.

As the railway gun assumed more importance on the Western Front, the tendency to get bigger and better was hard to resist. The 12-inch guns had turned out to be a practical proposition, and the French were making even larger calibers, so it seemed worthwhile to consider another increase.

Just prior to the war the Japanese Government had ordered some 14-inch naval guns from the Elswick Ordnance Company, and the last two guns of this contract had never been delivered due to the intervention of the war. The company now approached the War Office offering the two guns as potentials for railway mounting. The offer was accepted, and Elswick were asked to produce suitable mountings.

It turned out, of course, that these were the last railway mountings ever to be designed and built for the British Army, though nobody would have thought it likely at the time. The design was simple and robust, and until

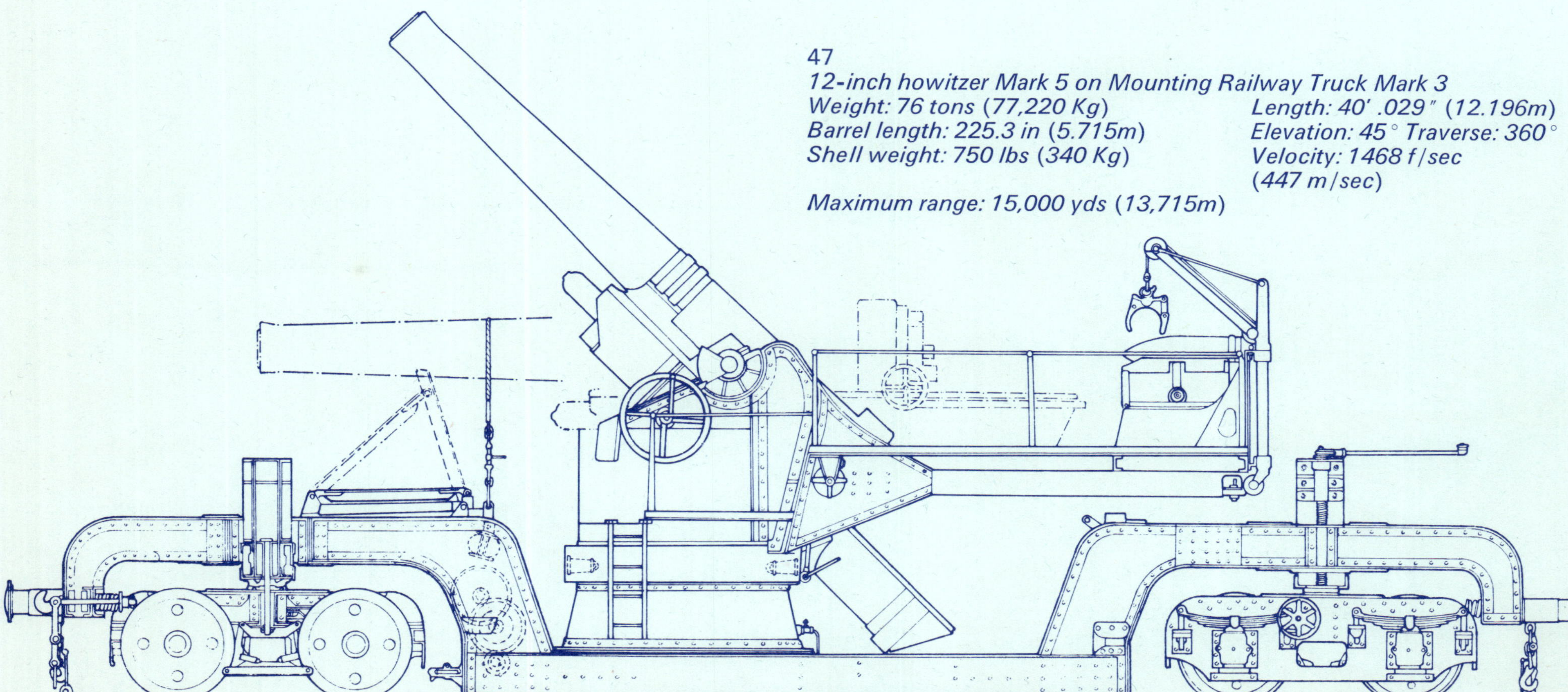

47
12-inch howitzer Mark 5 on Mounting Railway Truck Mark 3
Weight: 76 tons (77,220 Kg) — *Length: 40′ .029″ (12.196m)*
Barrel length: 225.3 in (5.715m) — *Elevation: 45° Traverse: 360°*
Shell weight: 750 lbs (340 Kg) — *Velocity: 1468 f/sec (447 m/sec)*
Maximum range: 15,000 yds (13,715m)

the development of the German guns of the late 1930s was probably among the world's best. It was a simple box structure suspended on two sets of bogies at each end. The gun was slipper-mounted to a cradle trunnioned to the side members. Four degrees of traverse were allowed by warping the end of the mounting across the front bogie bolster, so that in order to get a reasonable arc of fire a curved track had to be laid. A thirty-horsepower Siddeley engine drove a generator to provide power for the ammunition hoists, but the remainder of the operation of the gun was by hand.

The guns themselves were a 'handed' pair, originally intended for side-by-side mounting in a ship's turret, and consequently one exhibited an unusual feature for a British Army gun, a breech block which opened to the left. But, by way of a digression, it might be said here that all the early 14-inch guns in British Service were odd in some respect. The
45

British 12-inch howitzers Mark 3 in action in France. The gunlayer is loading the gun preparatory to firing, while the next shell has just been hoisted inboard.

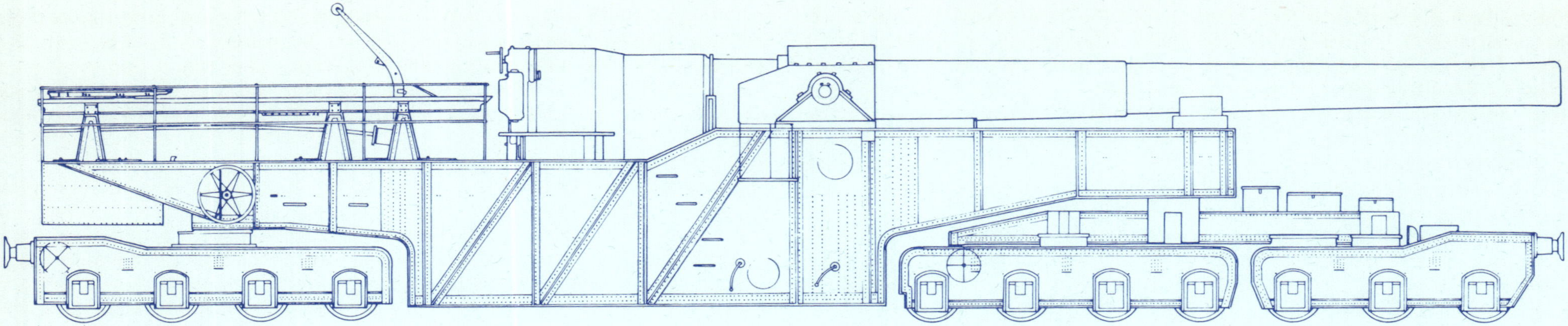

48
12-inch Gun Mark 9 on Mounting Railway Truck Mark 2 (Elswick Ordnance Company design)
Weight: 171.15 tons (173,736 Kg)
Barrel length: 496.5 in (12.598m)
Shell weight: 850 lbs (385 Kg)
Length: 70′ 8″ (21.44m)
Elevation: 36° Traverse: 1°
Velocity: 2525 f/sec (770 m/sec)
Maximum range: 28,200 yds (25,785m)

The front end of a Vickers pattern 12-inch gun. This photograph is particularly interesting in view of the two right-hand figures. The man with the telescope is a Royal Garrison Artillery signaller, while the man behind him is a Royal Flying Corps Officer. The remainder of the personnel are of 53 Siege Battery, RGA. It is believed that this must be one of the earlist photographs to indicate the co-operation of the RFC with the heavy guns for aerial observation of long-range fire.

Mark 1 type had been made by Elswick for the Chilean Navy; the Mark 2s had been made by the Bethlehem Steel Corporation in America for a Greek battle cruiser being made in Germany, and were 'appropriated' en route by the Royal Navy who used them to arm monitors. So the addition of a brace of Japanese weapons raised few eyebrows. Later, other 14-inch guns of Marks 1, 1+ and 6 were authorized for modification to suit railway mountings, but so far as is known none were actually fitted, and the Mark 3 guns were the only ones which ever entered service.

The idea of these guns had been born in late 1916, but it was the beginning of 1918 before they were completed. They were then shipped to France and arrived there in May 1918, being issued to 471 Siege Battery, RA. One gun, christened 'HMG. Boche-Buster' and commanded by Major S. M. Cleeve, RA, was sent to the Arras area, the other 'HMG. Scene-Shifter' under Captain M. B. Elderton RA went to the Bethune area. 'Boche-Buster' was distinguished, on 8 August 1918, by a visit from His Majesty King George V, while the gun was in position at Mareuil. Under His Majesty's supervision the gun opened fire on Douai railway station at a range of 18 miles and scored a direct hit with the first shot, a round which went into regimental history as the 'King's Shot'.

A common practice with towed artillery is the production of a mounting which will accept either a gun or a larger-caliber howitzer. This philosophy was applied to the 14-inch mounting and a suitable 18-inch howitzer was asked for in 1917. At that time no weapon of this caliber existed, so there could be no utilizing a naval reserve barrel or 'appropriating' a foreign-order barrel. It would have to be designed and made from scratch. But since a howitzer is less severely stressed than a gun, it was possible to build the piece to the same external dimensions as the 14-inch gun, notwithstanding the larger caliber. The only major alteration was in the breech mechanism, necessarily larger and some two and a half tons heavier.

The 18-inch was not completed until after the war had ended, being officially introduced into service in 1918. Two barrels were constructed and mounted, in turn, on a 14-inch mounting for proof firing, after which they were dismounted and stored. In 1926 the two 14-inch guns were declared obsolete, and instead of replacing them with barrels of the same caliber but of different Mark, the two 18-inch barrels were mounted in their stead. During the rest of the interwar years one equipment was fired once or twice on Salisbury Plain, but most of their time was spent in store.

In 1940, with every available gun in Britain being turned out to defend the country against possible invasion, the two 18-inch howitzers were sent down to Dover. It is not clear what they were expected to do there, since their range was too short for cross-Channel bombardment, and their mounting was ill-suited to use in the coast defense role. But they looked impressive in the newsreels and in newspaper photographs, so they doubtless paid for their keep by sustaining morale at a difficult time. Evidence points to the intention of using them in France had the 'phoney war' continued, since a design of replacement barrel was approved on 6 March 1940, but four days later the German Army squashed that idea, and the barrels were never made.

In 1944 there was a possibility that the howitzers might find useful occupation in the invasion of Europe. It was intended that they should be shipped to France once a port was secured, and then moved up to batter the concrete defenses anticipated on the German frontier. (It will be recalled that the US 14-inch M1920 was similarly earmarked at this time.) An anti-concrete shell was designed and approved in April 1944, but as the invasion plans became more detailed, and more information on German defense works came to light, there seemed to be less and less reason to employ such ponderous weapons. After the Allied air forces had finished with the French railway system, the possibility of moving 250 tons along them also appeared doubtful, and eventually the idea was dropped. When the war ended the guns were withdrawn, and in April 1947 they were officially declared obsolete.

It might be as well, at this juncture, to dispose of the 18-inch gun myth. From time to time queries are raised about a British 18-inch railway gun, but this is no more than layman's confusion between gun and howitzer. An 18-inch gun was indeed designed for railway mounting, in order to provide a very long-range weapon, and the barrel was built and proved. But by that time the postwar reaction had set in and the demand was no longer acceptable. The gun barrel was taken over and used for trials which culminated in the development of an entirely new 18-inch gun for naval use. At the same time there was discussion about producing a 20-inch naval gun, and, inevitably, the question of railway mounting it also came up. But the Washington Conference knocked the 20-inch on its head and no more was heard of it. To sum up—the 18-inch gun was never a British Land Service weapon; the 18-inch howitzer was the largest Land Service weapon ever issued.

Before leaving British railway artillery, two designs which never reached the hardware stage might well be mentioned. During the immediate postwar years the Paris Gun and its almost incredible range led to an orgy of drawing by almost every government and gunmaker in the world as they strove to emulate the Krupp design. The Elswick Company had made its own experiments with high velocity guns well before the war and appreciated most of the difficulties involved, and in the 1920s they produced a number of designs of long-range railway guns of 24cm caliber. These were all based on a solid ground platform onto which the mounting could be anchored, and all used long barrels supported by bracing in order to achieve the necessary velocity. Although a design of long range gun was flirted with by the British War Office in late 1918. It was done largely to satisfy Parliament that such a weapon was within the capability of the Allies and not entirely a German patent, much as the French 52cm howitzer had been produced to outstrip the German 42cm model. But the army did not really need such a gun, and once the war was over the design was quietly dropped and Elswick were unable to rekindle any interest with their 24cm designs.

The second design is one of the few railway gun designs ever to emanate from an official department rather than from a private gunmaker. The six-inch howitzer of 26 hundredweights was developed during the First World War as a replacement for the six-inch of 30 hundredweights, in order to provide field armies with a reasonably portable heavy howitzer. In its normal field role it was mounted on a two-wheeled box trail type of carriage for traction either by horse draught or by motor vehicle, later models being fitted with pneumatic tires. But at some time during the 1930s—the records are far from clear on this point—a demand arose for a lightweight railway gun. At that time the railway weapons remaining in British service were the 9.2-inch Mark 13 gun, the 12-inch howitzers Marks 3 and 5 and the 18-inch howitzer, a collection ranging from 60 to 250 tons in weight. It appears that an opinion arose that a lighter rail weapon might be of value in situations where the available trackage and bridges were insufficient to carry the bigger weapons. Since the six-inch 26 hundredweight howitzer was about the lightest weapon available with a

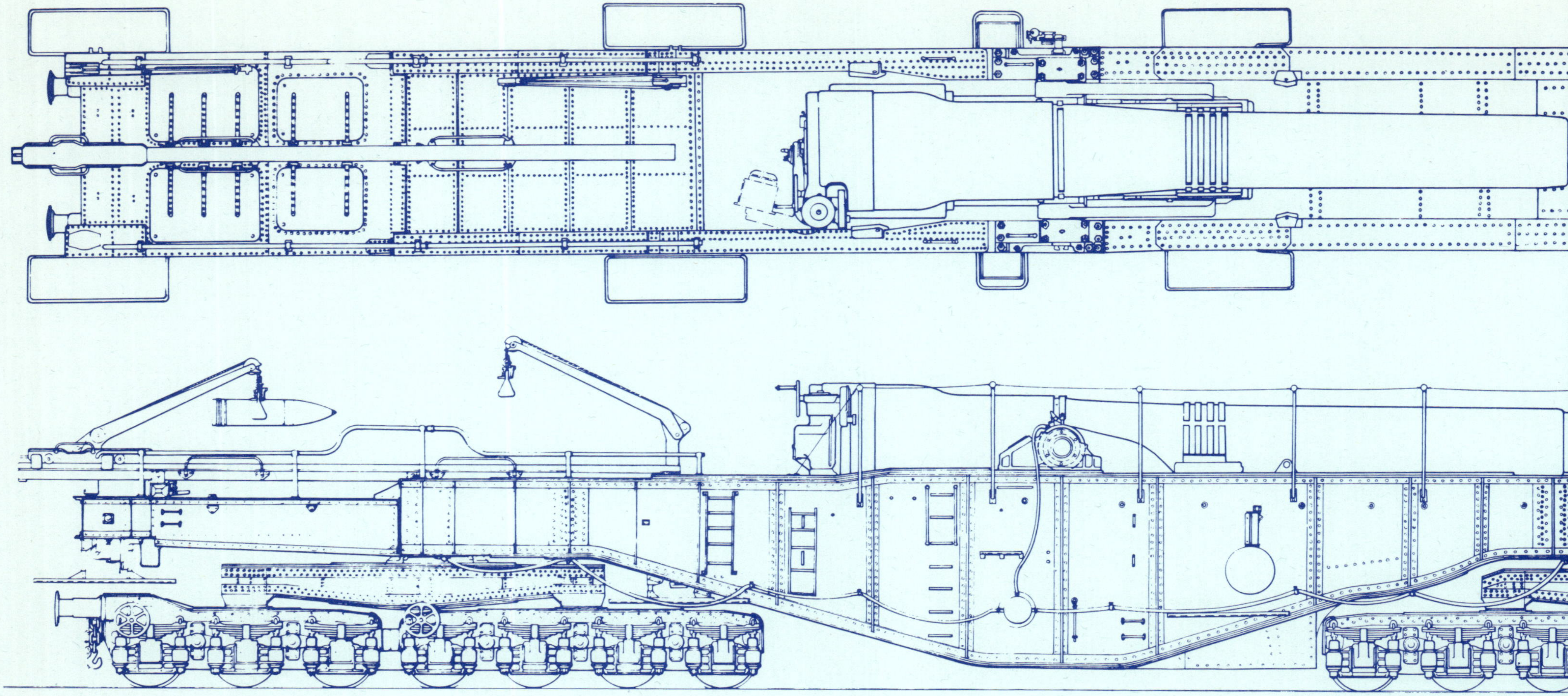

49
14-inch Gun Mark 3 on Mounting Railway Truck Mark 1
Weight: 248 tons (251,968 Kg) *Length: 87′ 4″ (26.618m)*
Barrel length: 648.5 in (16.459m) *Elevation: 40° Traverse: 2°*
Shell weight: 1400 lbs (635 Kg) or 1586 lbs (719 Kg)
Velocity: 2600 f/sec (792 m/sec) or 2450 f/sec (746 m/sec)
Maximum range: 38,000 yds (34,747m) or 34,006 yds (31,640m)
18-inch howitzer on the same mounting :
Weight: 250.4 tons (254.406 Kg) *Length: 87′ 4″ (26.618m)*
Barrel length: 648.5 in (16.459m) *Elevation 40° Traverse: 2°*
Shell weight: 2500 lbs (1134 Kg) *Velocity: 1880 f/sec (573 m/sec)*
Maximum range: 22,300 yds (20,391m)

53
A proposal for a long-range gun train, drawn by a diesel-electric locomotive. A spacer truck takes care of the barrel overhang, followed by three ammunition and store trucks and a personnel coach.

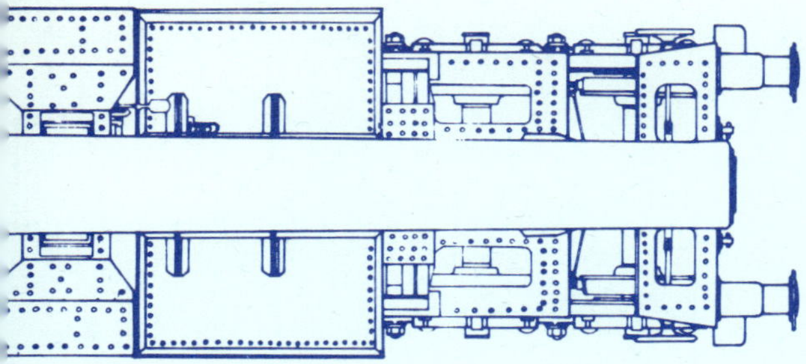

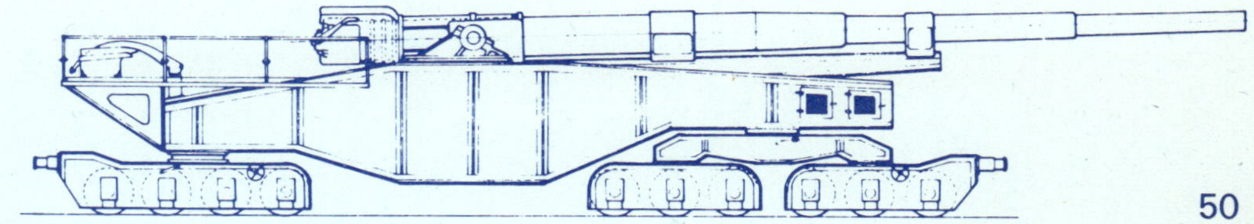

50

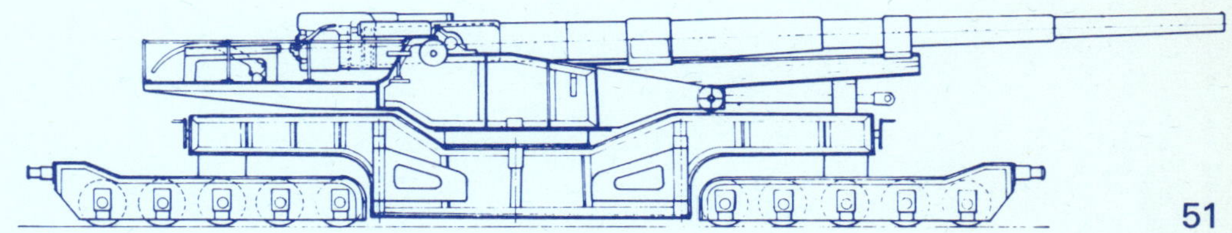

51

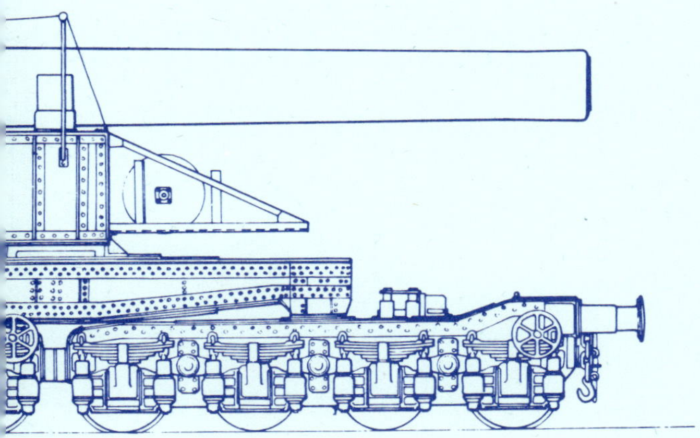

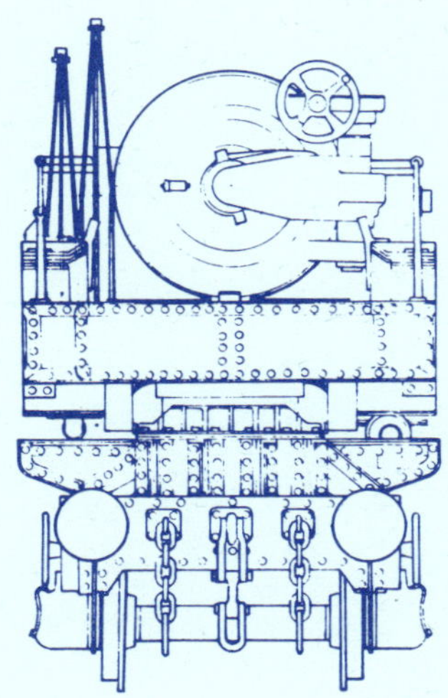

50
One of the Armstrong designs for a long-range gun, developed in the early 1920s after the effect of the Paris Gun had surprised the world.

51
Another Armstrong long-range gun design on a fully traversing mounting. It is interesting to see the extra length of cradle employed in these designs as an alternative to bracing the barrel for support.

52
A three-bogie Armstrong design of long-range gun with an unusual method of raising the gun barrel. The drawings show the travelling position with the ground platform details, and also the firing position, the hydraulic lifting system being used to control gun elevation and the mounting lowered onto the platform with the bogies removed.

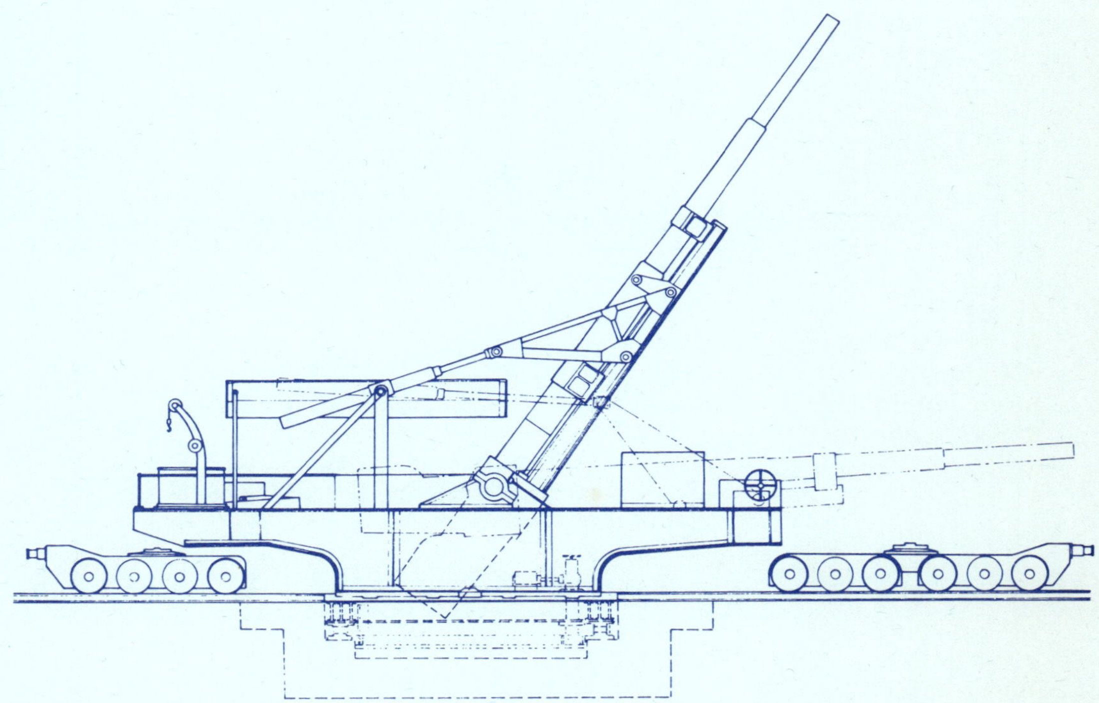

52

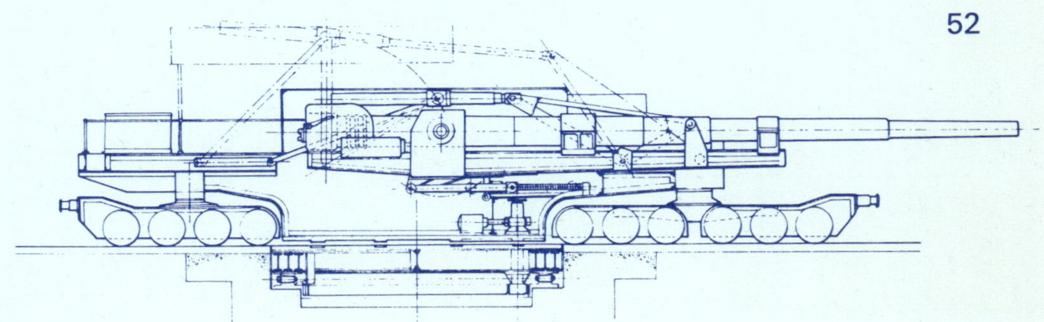

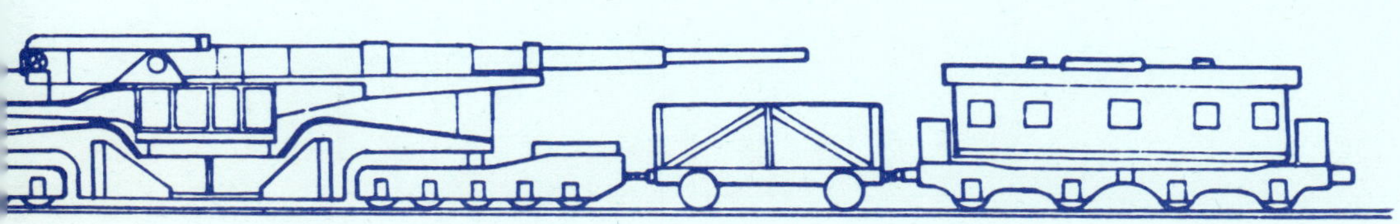

The British 18-inch howitzer fires, strangely deserted except for a pair of disembodied legs showing beneath the mounting.

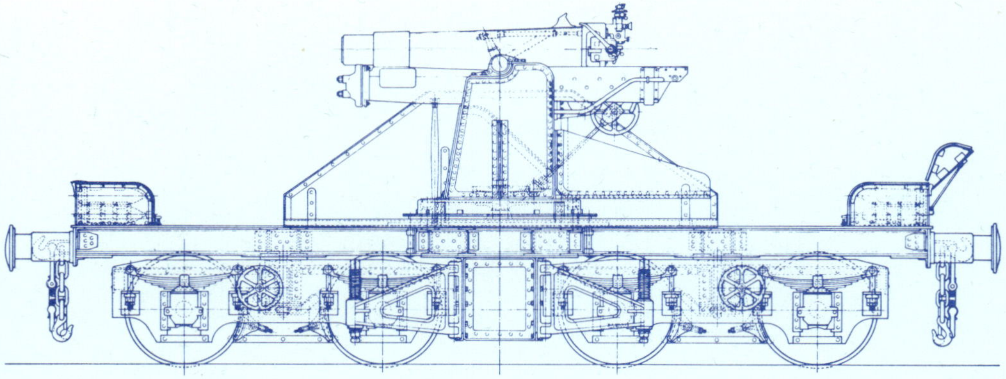

54
The British six-inch 26 hundredweight howitzer railway mounting design which never reached service.

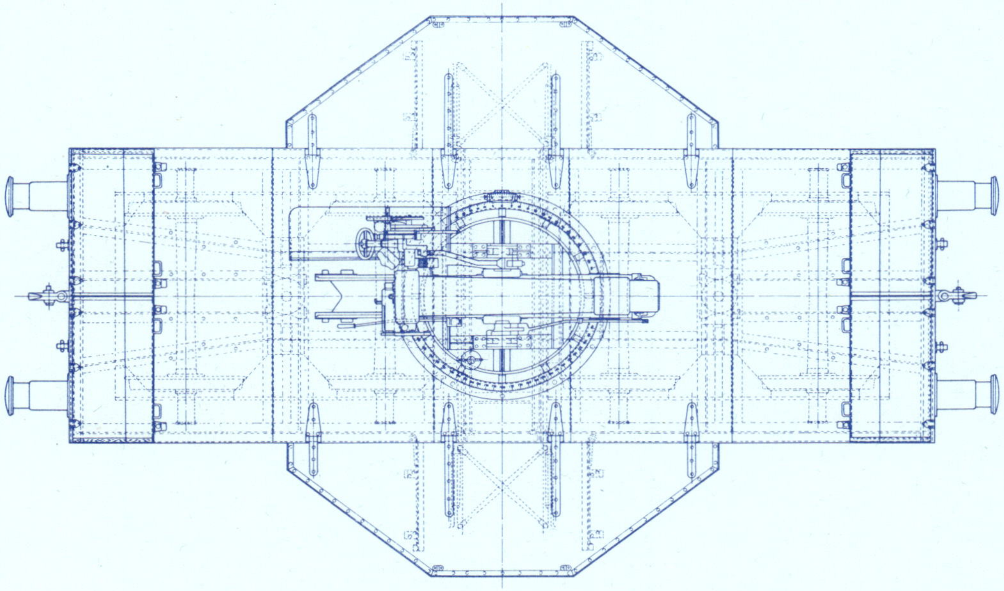

respectable shell power, it was selected for the task, and a suitable mounting designed.

It will be seen from the drawing that the truck has a very short wheelbase, allowing it to operate on tight radius curves, and the howitzer is assembled to a simple top carriage revolving on a turntable on the wagon bed. In the field role this howitzer was provided with a quick-loading gear which allowed the barrel to be disconnected from the elevating and sighting gear and swung to the horizontal for loading, without interfering with the sight setting and gunlaying operations, but this facility appears to have been omitted from the rail version. Short outriggers with screw jacks were provided, two on each side, and these were used in conjunction with pieces of one-foot-square wood skidding to provide a firm firing base, and relieve the truck suspension of the firing shock. In common with British freight stock practice of the time no train brakes were fitted, each bogie having a hand-operated brake.

It is, however, questionable that any of these weapons were ever built, beyond (perhaps) a prototype. There is no record in 'List of Changes in War Material and Stores' of the pattern ever being introduced, and the only available drawings are not marked 'SEALED' as would have been the case had the design been formally accepted for service use. It is highly likely that once the design had been drawn up it was shelved due to the greater priority of more vital weapons in the rearmament program of the time. Other evidence available points to the fact that the design was re-considered in 1940 as a possible Home Defense weapon. Its lightness and short wheelbase would have made it a useful reserve weapon which could have been held at strategic rail junctions and rapidly entrained to threatened points on the coast using lighter trackage which would have been unable to carry the heavier weapons. And, of course, the heavier weapons were few in number and earmarked for specific positions on the coast already.

Once again, though, pressure of events denied the six-inch 26-hundredweight railway gun its place in the Vocabulary of Ordnance Stores. In 1940 the losses of equipment in France had been so great that every Ordnance Factory and gunmaker was working full blast trying to make sufficient of the standard issue items to re-equip the army and provide for future expansion, and facilities for manufacturing a fresh design were simply not available. Indeed, weapons which had been fully approved and which were urgently needed found it hard to locate a maker—witness the delay in supplying the Army with the six-pounder anti-tank gun at this time. The six-inch 26-hundredweight railway howitzer was finally turned down on this score and the design completely abandoned. Had a similar design been required at a later date it is doubtful whether this one would have been considered, since the howitzer itself had by that time been replaced in service by the 5.5-inch gun-howitzer, a weapon with greater range and flexibility, and it would not have been possible to drop this weapon into the same mounting. A complete redesign would have been called for, and by that time the war had moved into a more fluid phase in which railway guns had no place.

The 18-inch howitzer ready for traveling. The metal framing is for sheeting the muzzle end with tarpaulins to disguise the shape from aerial recognition.

German Railway Guns

Germany, surrounded on almost all sides and with an intensive railway network was an ideal environment for railway gun employment. It is all the more remarkable that there seems to have been no attempt at developing such weapons prior to 1914. It may well be that Krupp's produced some designs, but there is no record of their exhibiting at any of the many international exhibitions of the time, and certainly the German Army showed no interest. The nearest application seems to have been Krupp's first 42cm howitzer design which was transported by special rail wagons but assembled on a ground platform for firing, and even that was turned down by the army in favor of a lighter road-mobile version. However, once the First World War began, railway guns were soon in demand, and a wide variety of designs were produced. Like Britain and France, the majority of the ordnance came from naval reserve stocks, since except for the 42cm siege howitzers Germany was little better prepared in the heavy artillery line than anyone else.

The smallest German railway gun was an extemporization of a sort not attempted by anyone else. This was the 17cm gun *Samuel*, which was no more than the standard 17cm heavy field gun tied down into a railway truck. It could be traversed by pulling the trail to one side or another of the truck, giving ten degrees of movement either side of the center line, plus, of course, the normal 12 degrees of traverse of the gun on its carriage. The origin of the name 'Samuel' is unknown, but it will be seen that the christening of railway guns with distinctive names was a common German practice.

Among the most numerous of the German guns was the 21cm *Peter Adalbert*. As with the majority of German wartime designs, there was none of the makeshift appearances of some contemporary British or French designs about this. It was of conventional girder box construction, braced by transoms and carried on bogies, and was indeed revived in the Second World War construction programs with hardly any alteration. The gun was cradle mounted with a hydro-pneumatic recoil system to absorb most of the firing stress, the rest being taken up in the rolling action of the mounting.

The 28cm *Bruno* (a name to be revived in later years) was constructed around a gun removed from the reserve stocks of the von der Tann class battleships, and was officially known as 'EuB Gerüst', meaning 'Eisenbahn und Barbetten Gerüst' or 'Railway and Barbette Mounting', indicating its capability of being adapted to either railway or coast artillery mountings. The rail design was among the first to use an integral turntable. Beneath the mounting was a ball-bearing supported platform which could be

German 24cm gun of 1916 showing an unusual system of jacking the weight off the bogies with externally-mounted screw-jacks.

A German 28cm 'Bruno' of the First World War. Other photographs in the author's possession show this gun prominently labelled 'Captured by the British 4th Army.' Later photographs, this one included, show that 'British 4th Army' has been roughly erased, and 'AUS' painted in. A touch of bitterness creeping in here.

55

28cm Schiffskanone L/40 'Bruno' in Eisenbahnlafette

Weight: 121 tons (123,000 Kg)	*Length over buffers: 75' 0" (22.8m)*
Barrel length: 398.43 in (10.120m)	*Elevation: 45° Traverse: 21°*
Shell weight: 253.3 lbs (115 Kg);	*Velocity: 2560 f/sec (780 m/sec)*
Maximum range: 27,950 yds (25,580m)	

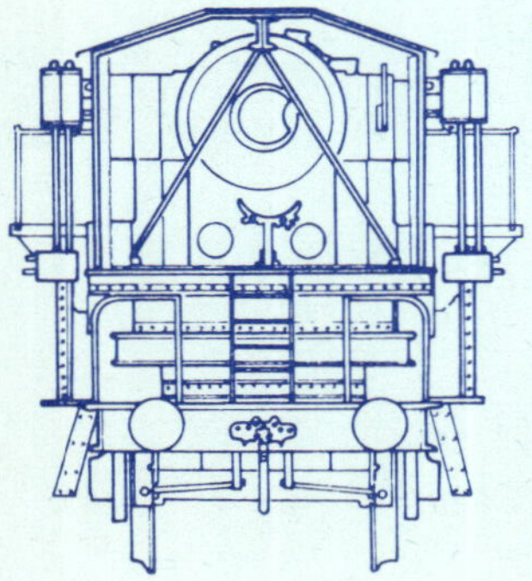

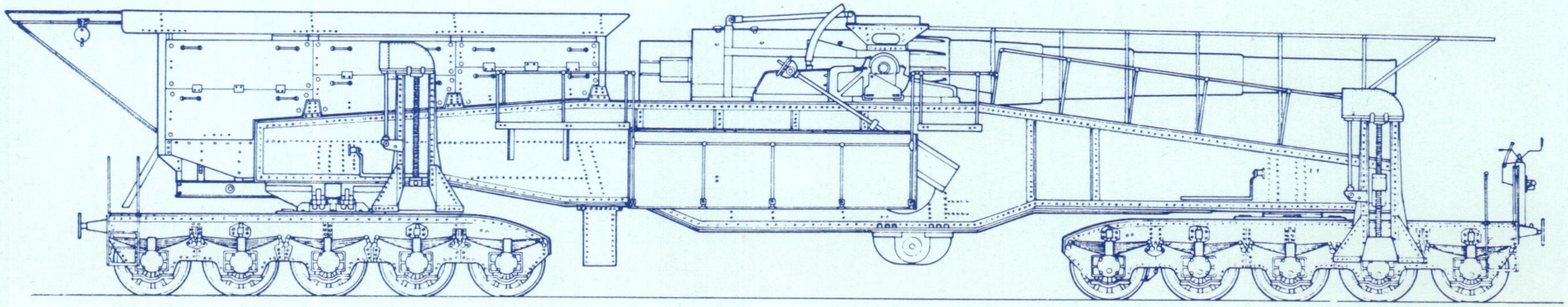

Small soldier, big shell. A kanonier tightening the fuze in a 1650 pound 38cm shell for the 'Max' gun in the background.

. . . Then a soldier,
Full of strange oaths and bearded like the pard,
Jealous in honor, sudden and quick in quarrel,
Seeking the bubble reputation,
Even in the cannon's mouth.'

(As you like it)
William Shakespeare

This cheery soul has a rope tied around his ankles and is about to be pulled through the gun to clean it. Or possibly not.

56
38cm Schiffskanone L/45 'Max' Eisenbahn Gerüst
Weight: 265.73 tons (270,000 Kg)
Barrel length: 671.26 in (17.050m)
Shell weight: 1652 lbs (750 Kg)
Length: 137' 9" (41.98m)
Elevation: 18.5° Traverse: 4°
Velocity: 2625 f/sec (800 m/sec)

Maximum range: 26,245 yds (24,000m)

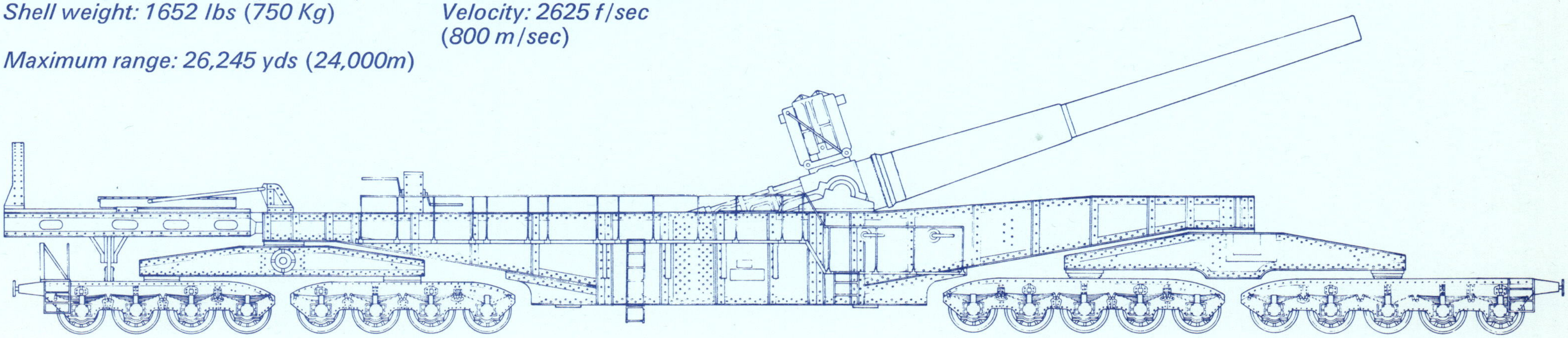

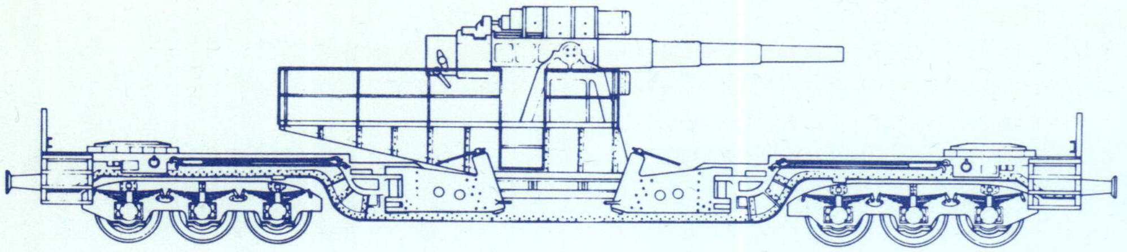

57
15cm Kanone in Eisenbahnlafette
Weight: 72.8 tons (74,000 Kg)
Barrel length: 234.6 in (5.690m)
Shell weight: 95 lbs (43 Kg)
Maximum range: 24,500 yds (22,500m)
Length over buffers: 65′ 11¼″ (20,100m)
Elevation: 45° Traverse: 360°
Velocity: 2640 f/sec (875 m/sec)

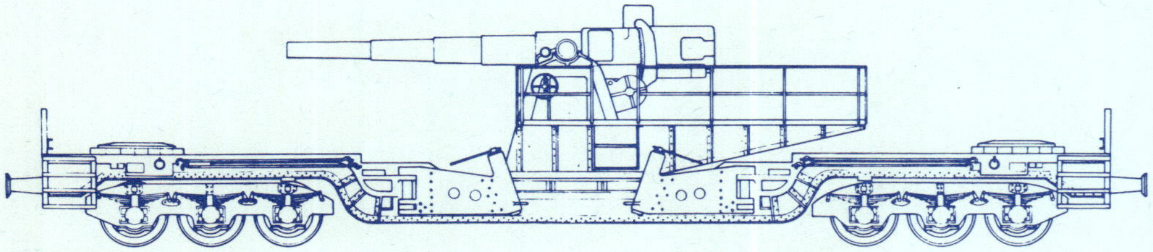

58
17cm Kanone in Eisenbahnlafette
Weight: 78.7 tons (80,000 Kg)
Barrel length: 271.6 in (6.900m)
Shell weight: 138 lbs (62.8 Kg)
Maximum range: 30,625 yds (27,200m)
Length over buffers: 65′ 11¼″ (20,100m)
Elevation: 45° Traverse: 360°
Velocity: 2870 f/sec (875 m/sec)

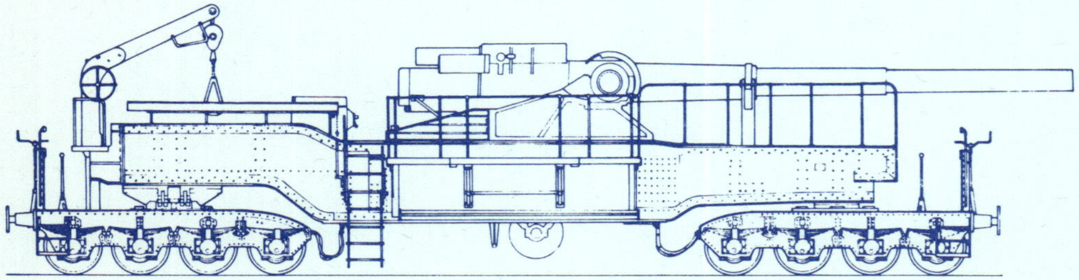

59
20.3cm Kanone in Eisenbahnlafette
Weight: 84.75 tons (86,100 Kg)
Barrel length: 478.34 in (12.150m)
Shell weight: 268 lbs (122 Kg)
Maximum range: 40,465 yds (37,000m)
Length over buffers: 63′ 9.55″ (19,445m)
Elevation: 47° Traverse: Nil
Velocity: 3035 f/sec (925 m/sec)

bolted down to a steel base dug into the ground. The gun was run across the prepared base on temporary track and jacked up until the bogies could be removed. Then the mounting was lowered until the lower section of the ball-race could be bolted to the ground platform. The bogies being withdrawn and the temporary track removed, the gun was now capable of 360 degrees of traverse on this central pivot with very little effort. It was estimated that the whole task of burying the ground platform and emplacing the gun took about one day's work, and a number of these guns were emplaced in this fashion along the Belgian coast in 1917–18 to act as coastal defense weapons.

The largest standard German railway gun of the First World War was the 38cm *Max*; this was also to be found either as a railway gun or as a concrete-emplaced coastal defense gun. Numbers of the latter type were also installed in Belgium, but the railway mountings were few in number and used on the Western Front, notably in the attacks on Verdun. A certain amount of confusion exists over the question of 38cm railway guns, since Dunkirk was frequently shelled by 38cm weapons, often described as Max guns. The author has fallen into this pit in the past, but recent researches show that two 38cm guns fired on Dunkirk at various times; one was a 38cm German Max in a Barbette mounting at Leugenboom, and the other was an Austrian 38cm of similar pattern known as 'Lulo'. So far as can be ascertained, no German 38cm railway gun ever fired on Dunkirk.

The Max barrel came from the reserves of the SMS *Bayern* and was mounted with cradle and recoil gear into a normal girder box type of mounting. This was normally placed on a ground platform for firing, though it was possible to treat the entire equipment as a rolling mount if time was short.

However, all other German guns of the First World War fade into insignificance when placed against the Paris Gun. This weapon caused consternation by bombarding Paris from a position in the Forest of Gobain, between Laon and Soissons, a range of approximately 68 miles. Between 23 March 1918 and 9 August a total of 303 shells landed in Paris, resulting in 256 people being killed and 620 wounded. A good deal of property damage was also done, but the German intention to strike terror into the hearts of the Parisians was not achieved, any more than their similar intent against the inhabitants of London in 1940.

The Paris Gun, or to give the weapon its correct title the 21cm Kaiser Wilhelm Geschutz, is generally attributed to Professor Rausenberger of Krupp's, one of the most brilliant gun designers ever to put pencil to paper, but it seems more likely that the original idea came from a Doktor von Eberhardt, an Austrian ballistician working with Krupp's. He did some arithmetic in 1916 and came to the conclusion that it should be possible to build a gun with a muzzle velocity high enough to put the shell into the stratosphere, so that the greater part of the trajectory would be virtually free of air resistance, and so develop a weapon with a range of 60 miles. He took this idea to Direktor Rausenberger, who confirmed his theories and set a team to work on the design. The next job was to find a customer; since the navy were more accustomed to dealing with heavy weapons, and since the basis of the design was to be a naval gun, Rausenberger took the designs to Admiral Rogge of the Naval Ordnance department. The navy approved of the idea, the army concurred, and work began. Then, at the end of 1916, Ludendorff threw a spanner in the works by suddenly demanding a range of 75 miles from the weapon. This caused a certain amount of uproar at Krupp's, even though they were used to the customers demanding more range and less weight with every succeeding design, but

The 21cm Kanone 12 with the upper carriage raised on its jacks into the firing position.

this was no mere whim on Ludendorff's part. He knew that when his planned reorganization of the front line was completed, a 60-mile range would be insufficient to reach Paris, the only worthwhile target, so 75 miles it had to be.

In spite of this setback, the first gun was built and proof fired at the Krupp testing ground at Meppen in the summer of 1917, and work was ordered to begin on the preparation of the firing site in September. Meanwhile the program suffered more delay when the original design of projectile was found to be unsuitable and fresh designs had to be worked out, made and test fired, and it was not until January 1918 that a serviceable shell was perfected. Eventually the guns were installed in March 1918 and opened fire forthwith. In spite of intensive counter-battery fire by the French, the guns were never damaged and were not silenced until the Allies advanced in August and took the area. Even then the guns were withdrawn in time and never fell into Allied hands; in fact, they vanished completely, never being found by any Allied Disarmament Commission in the postwar years.

The guns themselves—there were seven built—were based on the 38cm Max equipment, complete with its railway mounting. The barrel was bored out and lined with a 21cm tube, of which 42.3 feet projected from the muzzle of the 38cm barrel. To cover this, another reinforcing tube was shrunk on, and to the muzzle end a 19.7 foot smoothbore section was added. This 21cm liner was of such thickness that after it was worn beyond accuracy, it could be removed and re-bored, first to 24cm and a second time to 26cm caliber. The life of the gun at any one caliber was 50 rounds, and the erosion of the barrel was so severe that an outfit of serially numbered shells was prepared for each gun, each shell gradually increasing in diameter of the driving bands to compensate for wear. One of the guns was destroyed by a premature detonation of a shell during the bombardment, and it is believed that this was due to the inadvertent loading of an oversize shell out of its proper order.

The mounting was the normal pattern of side girders connected by transoms and carried on two double-bogie units. The gun was mounted in a ring cradle with a hydro-pneumatic recoil system which kept the recoil of this enormous weapon to no more than 50 inches. In order to balance the muzzle preponderance of the long barrel, a heavy counterweight was bolted to the top of the cradle.

The emplacement was a massive concrete bed some 35 feet in diameter

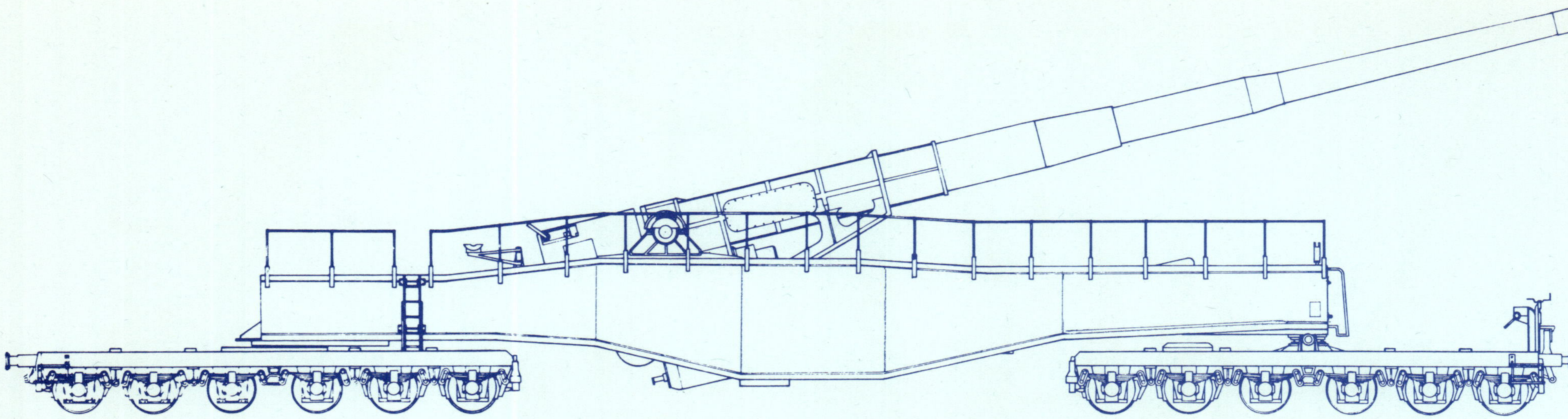

60
28cm Kanone 5 in Eisenbahnlafette
Weight: 215 tons (218,000 Kg)
Length overall, including barrel overhang: 69′ 8″ (21,234m)
Barrel length: 70′ 8″ (21.539m)
Elevation: 50° Traverse: 1°
Shell weight—standard: 561 lbs (255.5 Kg) Velocity: 3675 f/sec (1128 m/sec)
—rocket: 547 lbs (248.0 Kg) Velocity: 3710 f/sec (1130 m/sec)
—Arrow: 300 lbs (136.0 Kg) Velocity: 5000 f/sec (1524 m/sec)
Maximum ranges—standard shell: 68,000 yds (62.18 Km)
—rocket-assisted shell: 94,600 yds (86.5 Km)
—Arrow shell: 93.8 miles (151 Km)

▷
A close view of the breech of the 21cm K 12 when in the firing position, showing why it was necessary to lift the entire mounting to ensure that the recoiling parts did not strike the ground. The gunner on the left is attaching the firing lanyard to the percussion firing mechanism, while the gun layer on the mounting makes final adjustments. This appears to be a practice drill period, since the wrapping on the barrel, to protect the polished bearing surfaces, has not been removed.
An unusual picture of the K 12 gun being moved down a spur into position, while raised to the firing height. This illustrates several details of the mounting, particularly the massive casting on the front bogie which forms the link to the controlled carriage recoil system when firing from a turntable. ▷▷

61
21cm Kanone 12 in Eisenbahnlafette
Weight: 297 tons (302,000 Kg)
Length over buffers: 135′ 6″ (41,300m)
Barrel length: 109′ 3″ (33.3m)
Elevation: 55° Traverse: 0.5°
Shell weight: 236.5 lbs (107.5 Kg)
Velocity: 4920 f/sec (1500 m/sec)
Maximum range: 71.5 miles (115 Km)

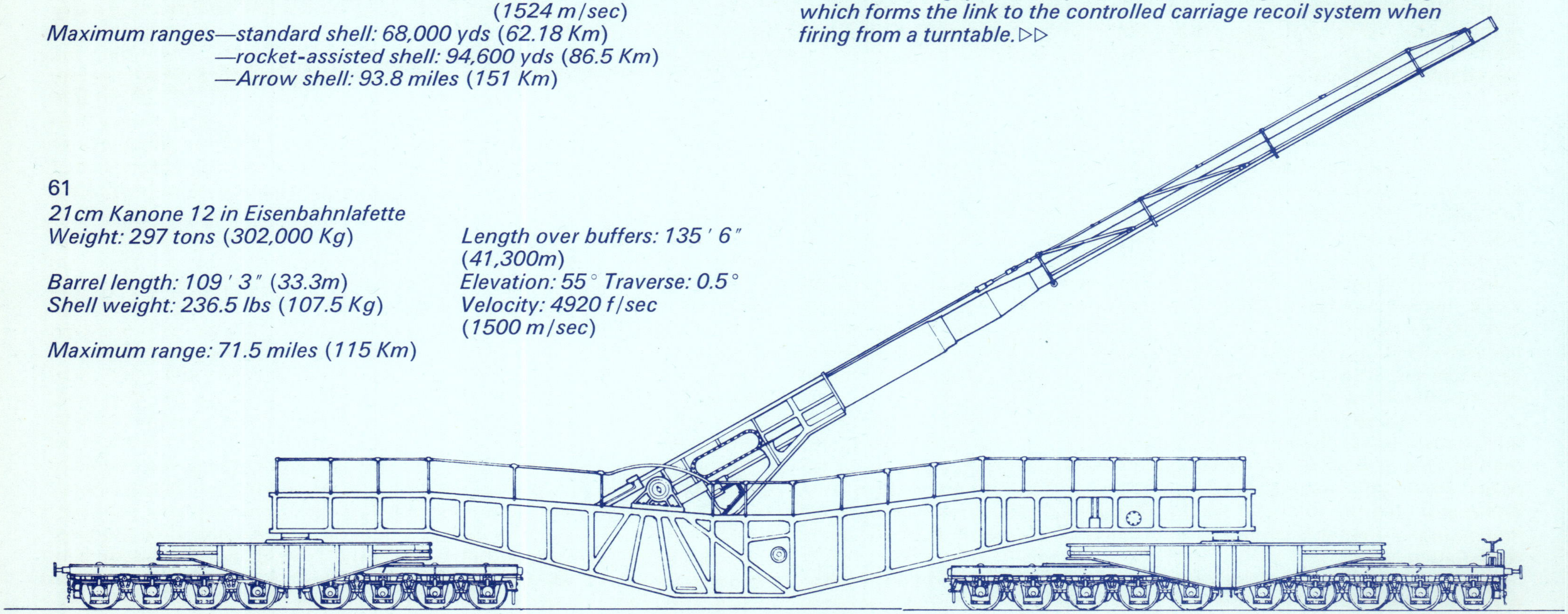

and 15 feet deep sunk into the ground, carrying a 28-foot diameter racer ring. On top of this went the rotating platform section, carried on 112 eight-inch steel balls. This unit was made up of heavy steel castings, raised at the sides and with a section of track running through a depressed portion in the center. To bring the gun into action it was run across this base and then raised by screw jacks until the bogie units could be disconnected and towed away. The upper section of the platform was then rotated through 90 degrees and the mounting lowered so as to rest on the raised outer sections, when key plates on the mounting engaged with corresponding plates on the platform and the mounting could be bolted to the fast base.

To fire, the gun was first loaded with the barrel in the horizontal position. The correct numbered shell was rammed, and the length of ram measured, since this increased gradually due to erosion. From this measurement the chamber volume was calculated, and from that dimension a weight of charge was worked out which would achieve the necessary range at a fixed elevation of 55 degrees. The charge was then weighed and bagged in the magazine, taken to the gun and loaded. The gun was then elevated and fired. The barrel was braced to prevent it drooping under its own weight, but even this failed to prevent it vibrating when fired, the movement being as much as a meter above and below its normal alignment and lasting for almost a minute after the shell had left.

In order to withstand the extreme accelerations involved, the shells were unlike anything which had been made before. A normal shell with a simple copper driving band would never stand up to the stress, and a special projectile with the walls milled into grooves to fit the gun rifling was used. This distributed the rotational stress through a greater area. In addition, a heavy copper driving band and another heavy copper centering band at the shell shoulder were also used, principally acting as gas seals to confine the propellant gas behind the shell and prevent leakage past the shell body, which not only meant wasted effort but also increased the rate of wear of the gun. There was nothing supernatural about the Paris Gun. It was an exercise in ballistics translated into reality on sound gunnery principles, and its shadow was to fall over gun designers for several years to come.

After the war the various Disarmament Commissions saw to it that every German railway gun went to the scrapyard, and when the Third Reich came into being, railway guns were high on the list of priorities. But during the lean years of the 1920s the designers had been quietly at work. In the late 1920s the German Army requested design work to begin on a 21cm gun capable of outranging the Paris Gun, since the fact that the navy had sponsored and manned that weapon sat somewhat hard on the backs of the soldiers. They were determined that one day the army would show that the sailors didn't have a monopoly in the big gun business. Another project which first went onto the drawing board at about this time was the Vogele portable turntable, which was to form a part of the equipment of many German railway guns during the Second World War.

The long range gun project became the 21cm Kanone 12; with the paper calculation done, the way seemed clear in 1933 to begin turning the drawings into hardware. The prime question to be resolved was whether improvements in materials had made it possible to use a normal type of shell with driving bands, or whether the complication of the rifled shell used with the Paris Gun would still have to be resorted to. The designers had improved this latter system into a method in which the gun was to be rifled with a small number of deep grooves, and the shells were formed with raised ribs. Two small-scale barrels were now made, one normally rifled and one rifled in the proposed way, and these were extensively tested. The deep-grooved barrel and ribbed shell proved to be the correct answer, and the full size 21cm barrel was accordingly built with eight rifling grooves.

The mounting was almost identical to the First World War 38cm Max design, a girder box carried on two twin bogie units, but a refinement was added. The mounting body was connected to the bogie bolsters by hydro-

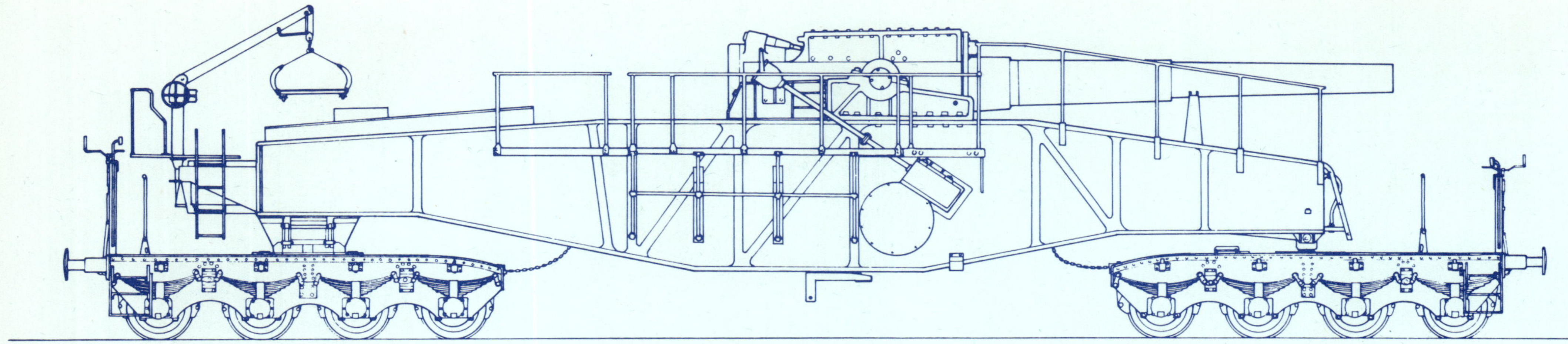

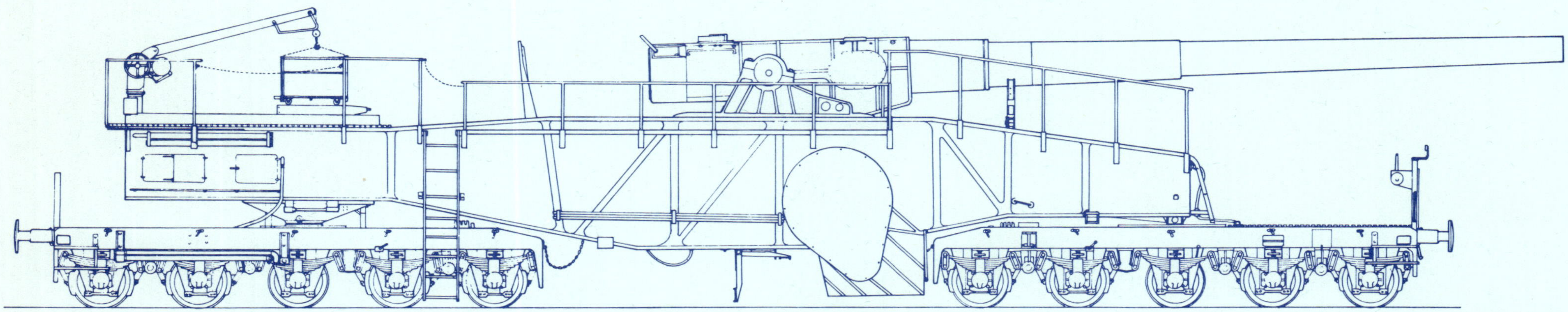

63
28cm Lange Bruno Kanone in Eisenbahnlafette
Weight: 121 tons (123,000 Kg)
Barrel length: 489.5 in (12.735m)
Shell weight: 625.8 lbs (284 Kg)
Maximum range: 39,500 yds (36,100m)
Length over buffers: 74 ′ 9.6″ (22.800m)
Elevation: 45° Traverse: 1°
Velocity: 2879 f/sec (875 m/sec)

64
28cm Schwere Bruno Kanone in Eisenbahnlafette
Weight: 116.13 tons (118,000 Kg)
Barrel length: 469.7 in (11.930m)
Shell weight: 625.8 lbs (284 Kg)
Maximum range: 39,050 yds (35,700m)
Length over buffers: 74 ′ 9.6″ (22,800m)
Elevation: 45° Traverse: 1°
Velocity: 2820 f/sec (860 m/sec)

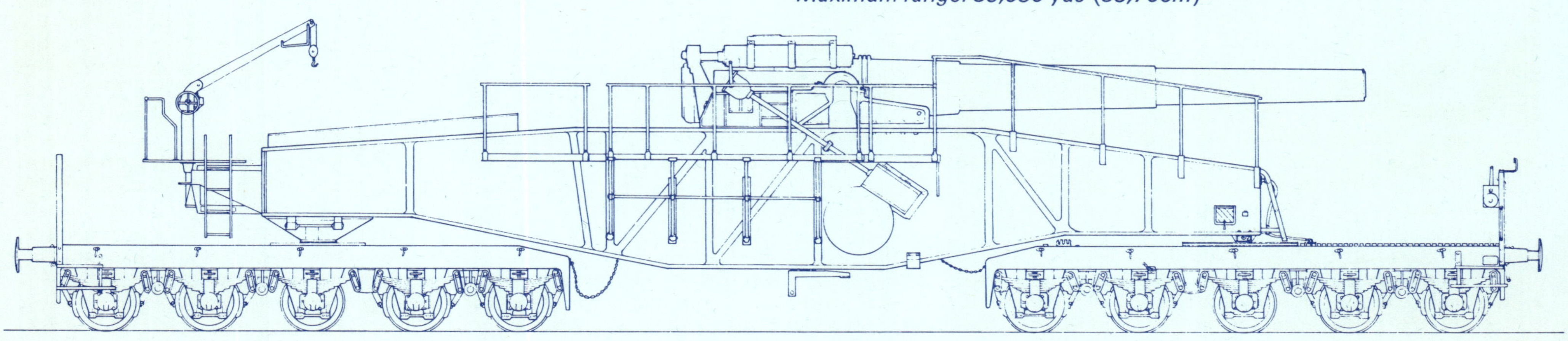

62◁

24cm Theodor Bruno Kanone in Eisenbahnlafette

Weight: 92.5 tons (94,000 Kg)

Barrel length: 330.7 in (8.400m)

Shell weight: 326.7 lbs (148.5 Kg)

Maximum range: 10,925 yds (10,000m)

Length over buffers: 68 ' 0" (20,700m)

Elevation: 45° Traverse: 1°

Velocity: 2215 f/sec (675 m/sec)

A 28cm 'Kurz Bruno' in action on the French coast. The gun is on a Vogele turntable and the camouflage has been rolled back on to the travelling shroud supports to allow the gun to be laid. Notice that there are small differences in the construction of this mounting from the general run of Kurz Bruno models, particularly in the system of external stiffening of the side panels.

pneumatic recoil systems. This allowed the whole structure to recoil across the bolsters for 98cm distance. The barrel was carried in a ring cradle with its own recoil system allowing a 65cm recoil of the barrel before the mounting system came into play. The effect of this dual system was to absorb all the recoil force and to place practically no thrust on the bogies.

To allow fine control of laying the front of the mounting could be traversed a small amount across the front bogie bolster, but basic pointing was done by the cross-track traverse system, the necessary track and turntable being carried in portable form in special wagons in the gun train. The front bogie unit could be driven along the cross track by electric motors controlled by the gunlayer.

The weapon was completed and issued to the army in March 1939, but, in the army's eyes, it had one glaring defect. Due to the problems involved in balancing the long barrel, the trunnions were set forward, and when the piece was elevated there was a danger of the breech striking the ground on recoil. To prevent this a hydraulic jacking system was built in which lifted the entire mounting structure to a height of one meter above the bolsters for firing. This had to be operated for every round, and lowered again to allow the gun to be re-loaded. The designers were asked to try again, and do away with this time-consuming performance. Some fundamental research into hydro-pneumatic balancing cylinders showed that it was possible to produce a system which would take the enormous weight involved, allow the trunnions to be moved back and yet still balance the gun, and so do away with the need for jacking. A second gun was built, incorporating this feature and also increasing the mounting recoil stroke across the bolsters to 1.5 meters for better stability, and this gun was issued in the summer of 1940. No more were ever made; while they were undoubtedly brilliant designs, they were hardly practical weapons. The two guns went down to the French coast late in 1940 and were used for some time as cross-channel bombardment guns, shelling Dover and Folkestone and many other parts of Kent until late 1941, but their subsequent fate is unknown.

As the design of the 21cm Kanone 12 went ahead, so did work on another gun, the 28cm Kanone 5, which was to become the backbone of the German long range artillery and one of the best railway guns ever made. Much of the early research work on the 21cm was drawn on, for a similar pattern of deep-grooved barrel was adopted, this time with 12 grooves. The mounting was the usual girder structure carried on two six-axle bogies, and no carriage recoil system was fitted. The equipment was provided with a portable turntable which could be quickly laid, and used 'controlled carriage recoil', a system in which the mounting front bogie was attached to the end of the turntable by a hydro-pneumatic recoil system, the piston rods of which were hooked to the bogie and the cylinders anchored to the turntable.

The basic gun, as issued in 1937, achieved a range of 38.7 miles with a 561-pound shell, and the most interesting aspect of this weapon is the development of special ammunition which went on in order to increase the range. The first design to be adopted was a rocket-assisted shell. The forward portion of the shell carried a solid-fuel rocket motor ignited by a time fuze when the shell was almost at the top of its trajectory. The blast passed down a central pipe and vented at the base of the shell and this additional boost increased the range to 53.75 miles, albeit at the expense of accuracy. Another drawback was that the space occupied by the rocket motor and blast pipe had to come from space normally occupied by high explosive, and so little payload was left that it hardly seemed worth the

Bringing a 28cm 'Kurz Bruno' gun into action.
The travelling shroud is taken off, then the shroud support framework is unbolted and removed. When ready to open fire, having removed the netting, the ammunition is brought from the ammunition car on a trolley running on rails to the edge of the turntable, where it is transferred to another trolley running around the turntable track.
The complete round, shell and cartridge, is secured to a small sub-trolley which rides on the larger ones and thus can be easily transferred from one to the other.
The complete round is pushed to the rear of the gun, into a position allowing the sub-trolley and its load to be hoisted up to the gun's loading platform.

effort of getting it there.

The other method of reaching for more range was developed at the Peenemünde Research Establishment. As well as their well known efforts in the missile field, a lot of fundamental research into fin stabilized projectiles of all types was done at Peenemünde, and they produced the 'Peenemünde Arrow Shell', a long, finned, dart-like projectile with a stabilizing ring around the waist which held it steady in the gun barrel and which was discarded once clear of the muzzle, leaving a streamlined dart flying through the air. Since the projectile was fin-stabilized there was no need for rifling, and a 31cm smooth bore barrel was designed to fit the 28cm mounting. Since the resistance in the barrel was less than with a rifled weapon, a heavier charge could be fired, and the result was a range of 93.8 miles; two of these 31cm guns were built and were used in action against the US 1st Army in the closing stages of the war. It is rather amusing that the 28cm K5, which was virtually the hack weapon—28 were in service by the war's end—finally managed to out-range—by 22 miles—the 21cm K12 which set out to be the long-range super gun!

The Kanone 5 was a popular weapon with the soldiers; they named it *Schlanke Bertha*. It was generally less popular with the troops on the other end; the GIs at Anzio named one the *Anzio Express* from the speed and regularity of its deliveries. Like most German equipment it was meticulously thought out and the gun was supplied complete with two trains which were models of their kind. The gun train consisted of locomotive, gun, air-conditioned magazine car, a diesel locomotive, two shell cars each carrying 113 shells, two cartridge cars, an equipment truck carrying various small stores and tools, an armory car with the crew's personal and local defense weapons, a kitchen and dining car, a flatcar with a blast-protected fire control room built on it, a flatcar carrying a quadruple 20mm anti-aircraft gun, and three coaches which formed the detachment's living quarters.

The second train, the turntable train, comprised locomotive, a car carrying the turntable sections, one carrying track segments, a crane, three ammunition cars, a stores and tool car, a flatcar with anti-aircraft gun, personnel coach, and nine flat wagons for carrying the motor vehicles allotted to the battery. The diesel locomotive on the first train was for switching and moving ammunition cars on the gun position, moving cranes about, and, if necessary, moving the gun from its firing position to a position of concealment. It was an Orenstein & Koppel Model C14, weighing 40 tons.

To bring the weapon into action a suitable flat area astride the track was selected. The diesel locomotive brought up the car carrying the turntable track segments and the crane, and the segments were unloaded and set out in a circle. The empty car was removed and the turntable car brought up. The turntable, which actually formed the body of the car, was then jacked up and the bogies removed, and the circle of track segments completed by fitting bridge pieces across the main track. The turntable platform was then lowered on its jacks until the end rollers rested on the circular track built up from the segments. Since the level of the turntable was about 70cm above the original track, an inclined frog about 40 feet long was now laid and the diesel locomotive pushed the gun up this and onto the turntable platform. The carriage recoil system was then connected between the front end of the turntable and the leading bogie, the frog track removed, and the gun was ready for action. The diesel engine then brought up the air conditioned magazine car with an outfit of ammunition. To lay the gun the turntable was motored round electrically until the gun was pointed in the right direction, whereupon four hydraulic jacks were lowered to steady the table and take the firing shock from the traversing rollers.

While these 21cm and 28cm weapons were being designed and built, the German Army became a bit restive. It was going to be very nice to have these super guns, but what about plain ordinary guns to be going on with? They knew all about the best being the enemy of the good, and eventually in 1936, managed to get official backing for what became known as the 'Emergency Program'. This called for a variety of weapons to be produced and delivered by the Spring of 1939. This demand did not upset Krupp's (who were the only builders of railway guns in Germany); they simply reached into their archives for the drawings of the First World War mountings with one hand, and into the stock list of naval reserve guns with the other. The same old trick was about to be performed.

A design of 15cm gun as part of the long-term equipment program had been simmering for some time, and these were rapidly completed and issued in 1937 so as to leave the assembly shops clear for the 'Emergency Program'. This 15cm Kanone was an efficient enough design of platform mounting with outriggers, but 73 tons of six-inch gun is hardly worth the trouble of getting into action, and after taking 18 weapons the Army dropped the design. A 17cm gun was then produced to fit the same mounting, with the intention of replacing all the 15cm models, but it was still too small to be worthwhile and no more than six were ever built.

To turn now to the 'Emergency Program', the first design was the 24cm *Theodor Bruno* gun, an ex-coast artillery weapon dating from 1910. The mounting was a simple rolling pattern, adapted from a First World War design, and the gun was fitted to the mounting by an unusual design of rectangular cradle built up from steel castings and containing the recoil system cylinders. A small amount of traverse was provided by moving the rear end of the mounting across its bogie bolster by a hand-operated screw gear.

The next weapon was the 24cm gun *Theodor*, another ex-naval gun on a mounting almost identical with the *Theodor Bruno* model. With a supply of 24cm guns assured, Krupp's now turned to the 28cm caliber and produced four weapons, differing principally in nothing more than barrel dimensions and christened accordingly. The first, issued in 1937, was the 28cm *Kurz Bruno*, an ex-naval gun on a rolling mount which was more or less an enlarged and strengthened version of those made for the 24cm guns. The next one had a longer barrel and was called 28cm *Lange Bruno*. Three of these were built, and then came one with a short and heavy barrel which was called *Schwere Bruno*. This used up two remaining coast gun barrels of an old design which Krupp's just happened to have about the place.

65
28cm Bruno Neue Kanone in Eisenbahnlafette
Weight: 147.6 tons (150,000 Kg)
Length over buffers: 81' 9" (24,880m)
Barrel length: 645.67 in (16.400m)
Elevation: 50° Traverse: 1°
Shell weight: 582 lbs (265 Kg)
Velocity: 3264 f/sec (995 m/sec)
Maximum range: 51,000 yds (46,650m)

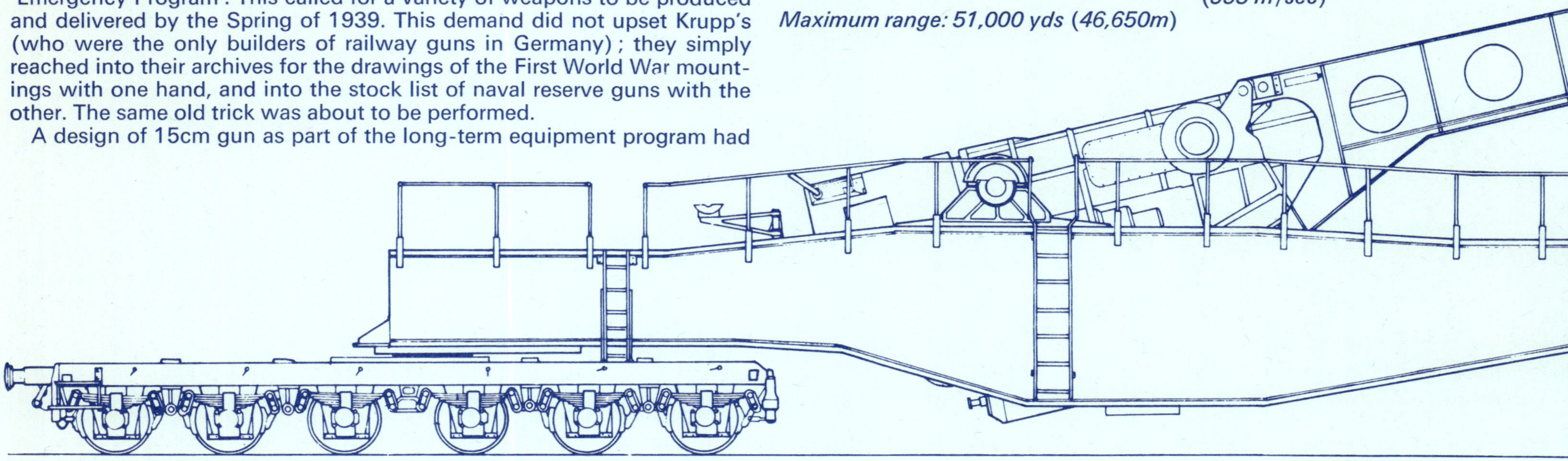

While these designs were being put together the army looked at their potential; the greatest range achieved by any of them was slightly more than 22 miles, and so the army asked if it might be possible to improve this figure somehow. Krupp's pointed out that if the army insisted on its 1939 delivery date, nothing could be done in the time, since it would mean designing a new barrel. Accepting this, the army then asked for a new weapon to be designed, leaving the delivery date open. Because of its being a new design, this one became *Bruno Neue* and was a modern 58-caliber weapon, giving a range of 28.75 miles. The mounting, while of the usual box type, was a much cleaner design and was provided with

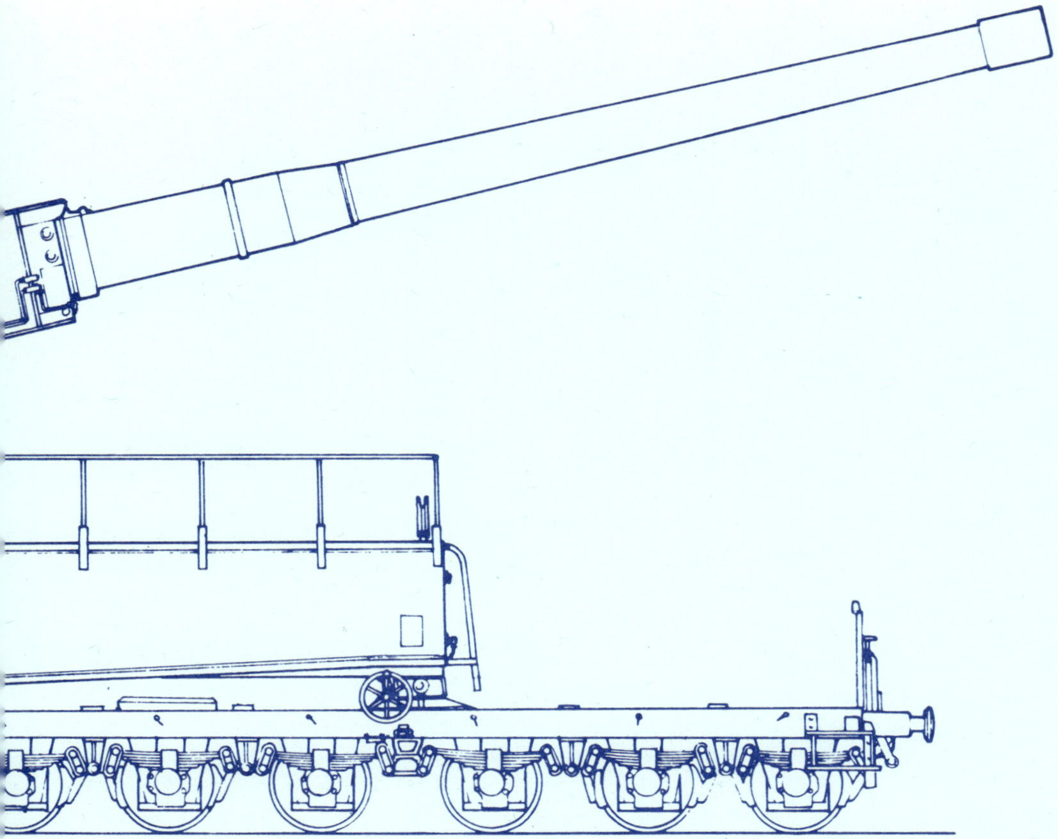

66

38cm Kanone 'Siegfried' in Eisenbahnlafette

Weight: 289.13 tons (294,000 Kg)

Barrel length: 64 ' 4¾ " (19.630m)

Shell weight: 1091 lbs (495 Kg)

Maximum range: 61,000 yds (55,700m)

Length over buffers: 112 ' 2½ " (31,320m)

Elevation: 45¾ ° Traverse: Nil

Velocity: 3445 f/sec (1050 m/sec)

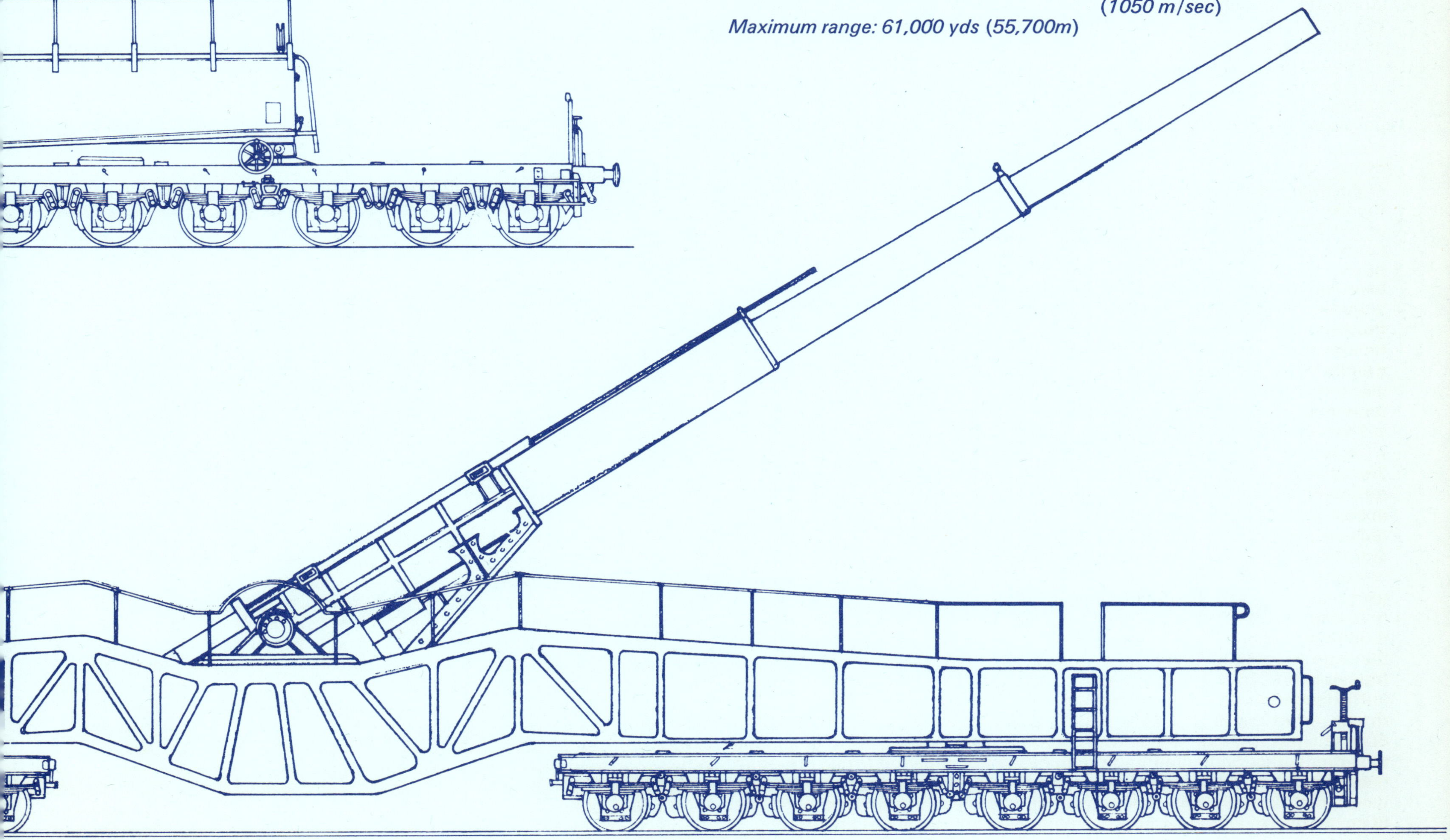

power assistance for elevation and loading. A portable turntable was provided with each gun, together with a controlled carriage recoil coupling. It was intended to make a large number of these, but production was stopped after only three had been built. The gun suffered from an unaccountable and hazardous ballistic irregularity which led to high bore pressures and erratic shooting; moreover the 28cm K5 design had proved to be so outstanding that the army preferred to see the work and material put into the production of more K5s.

In 1939, while the 'program' was under way, Krupp's were building a number of 20.3cm naval guns for the 'Blücher' class of warships. Eight guns were surplus to naval requirements, the gun program having got ahead of the shipbuilders, and the army, still worried about their stocks of railway guns, persuaded the navy to part with them and Krupp's to build suitable mountings. By adapting the First World War design of Peter Adalbert mounting, they were able to produce a suitable equipment quite quickly, and with this project in hand the army fell out, feeling quite pleased at obtaining eight guns so easily.

They were soon back knocking on Krupp's door. It had suddenly dawned on them that they were about to get eight guns of a non-standard caliber. Nothing else in the army was 20.3cm, which introduced a logistic problem in providing and supplying ammunition. Could Krupp's please rework the barrels to 21cm caliber so that standard army ammunition could be used? No, they couldn't. It was too late; the guns were built and rifled. They were therefore completed in 20.3cm and delivered in 1940. Krupp's then set to work to build a set of eight spare barrels in 21cm caliber, so that as the 20.3cm models wore out they could be replaced. Four of these were ready in 1944 and the other four under construction when six of the eight guns were captured by the Allies in France. Further work on that program was abandoned.

By the end of 1938, with all the various designs going smoothly, and the army satisfied for the time being, Krupp's decided to develop something heavier, and began by designing a 38cm gun which became known as *Siegfried.* To save time a naval 38cm gun was selected as the ordnance, since this had already been adopted as a coast artillery weapon and a long range shell had been designed for it.

The mounting, of the rolling type, was the usual box girder on bogies, bearing some resemblance to the 38cm Max of the First World War. No carriage traverse was provided, the gun train carrying the usual turntable, and the barrel was braced against droop. The design was approved in 1939 but it was the summer of 1943 before the first was delivered and only three of the eight ordered were completed before the war ended.

The next design to be attempted was the 40.6cm *Adolf,* another ex-naval and coast artillery gun. This was to be installed in the same mounting as the *Siegfried,* but the design ran into trouble over getting it down to dimensions which would pass the loading gauge. It is believed that one was finally built, but evidence on the matter is scant and contradictory.

The last of the German Second World War designs was as great a shock to the rest of the world as the Paris Gun had been in 1918. This was the 80cm caliber *Gustav Gerat,* the largest gun ever built. This enormous weapon had its beginning in an enquiry by the army in 1935 as to what power of gun would be needed to defeat the fortifications of the Maginot Line. The Krupp design office produced hypothetical solutions for 70cm, 80cm and 100cm caliber guns, sent them off and forgot about them; soldiers are always asking gun designers silly questions. Nothing further happened until Hitler visited the works in 1936 and asked the same question; he was given the same figures. Gustav Krupp von Bohlen und Halbach, a shrewd judge, set his drawing office the task of designing the 80cm model without waiting for any signs of approval, and in 1937 produced the drawings to the army. They were approved and work began on building the gun.

Manufacture was not easy, and it was not possible to have it ready in the Spring of 1940, the originally planned delivery date. The barrel was finally completed and proof fired early in 1941 and a year later the complete equipment was ready. It then went east to shell Sebastopol during the siege; was removed and sent to Leningrad, but was unable to get into action before the Soviet Army advanced and raised the siege of that city. It made one more brief appearance near Warsaw in 1944 and then, like that other Krupp masterwork the Paris Gun, it vanished completely, never being captured or found in postwar searches.

Since the complete weapon weighed 1329 tons, it had to be transported in pieces on special trains. On arrival at the selected site, four tracks were laid, two for the gun and two outer tracks for a travelling crane used to assemble it. The bogie units were pushed into place on the center tracks and locked together side-by-side to form double units and the mounting was then built up piece by piece on top of them. The gun barrel, in two sections, was joined by a junction nut and assembled to its jacket and breech ring, inserted into its cradle and the whole unit lowered into the trunnions. The whole process of preparing the site and building the gun took three weeks and 1420 men. One unresolved question is why the weapon used railway bogies as its basis; with the performance which had to be gone through to erect it, it might just as easily have been built on a plain ground platform.

For some years after the war it was believed that two of these weapons had been built, named Gustav and Dora. However, subsequent research has shown that only one Gustav was made; the confusion arose in a typically soldier-like manner. Krupp's presented the gun to the army, and named it Gustav in honor of the man who started it going; the brutal and licentious soldiery preferred to call it, more familiarly, Dora, which accounts for the confusion.

In 1943, in the face of increasing air attacks, the German Army began a development of considerable potential. It was decided that a railway gun needed to be capable of movement across country in order to be able to detour around bombed track or broken bridges. The 28cm Kanone five was proposed as a basis for the weapon and the mounting was to be built with an integral platform beneath, so that the equipment could be lowered onto this, the bogies removed, and the gun left in position with a 360 degree traverse. Next it was proposed that the barrel be removed to a special transporter suspended between two modified 'Tiger' tank chassis, while another carried the breech mechanism and two more lifted the mounting between them. This caravan would then drive off and could either take the gun round the obstacle and re-assemble it on the track the other side or place it into action on the integral platform at any convenient site, with or without a railway track. The whole project was still in the planning stage when the war ended. It remained for the Americans to produce the ultimate in this line of thought when they unveiled the 28cm M65 gun, for all the world like a railway gun without bogies and suspended between two specially designed truck transporters.

This marked the beginning of the end for the railway gun. Until then it had been necessary to use rails to move heavy ordnance about, but now it seemed possible to do it with wheels or tracks and go anywhere. Then came the guided missile which delivered immense payloads over incredible distances and needed almost nothing in the way of heavy machinery to launch it. Improved systems of battlefield surveillance made it virtually impossible to conceal a man, let alone a 200-ton gun tied to the railway track; indeed it was this difficulty of concealment which eventually hastened the demise of the American 28cm gun itself.

The writing was on the wall, and cost-conscious armies took little time to read the message. By 1950 not a single Western nation had a railway gun left in its inventories. Russia had received a number of 17cm guns as reparations from Finland, and in the immediate postwar years there were reports that they were continuing development of long-range projectiles for captured German 28cm guns (though how they expected to improve on what the Germans had done is an open question). It is doubtful, however, that any railway guns exist in Russia either, such has been the advance in missile technology over the last twenty years. As far as I am aware, the only railway guns in existence today are a handful of captured German weapons in the Ordnance Museum at Aberdeen Proving Ground in the USA.

It is, of course, progress; but at the same time it is a pity. The rail gun was a delightful combination of every boy's love of railways and the naturally awesome presence of a big gun. Not to have seen one in the flesh is a deprivation and we trust that this short dissertation on the species will, in some way, compensate for this regrettable lack.

The biggest gun in the world; the 80cm (31.5-inch) 'Gustav' peers round a curve before opening fire on Sebastopol.

'Gustav' elevated and ready to fire on its trial shoot on the Rugenwald range in 1942. All records and drawings of this weapon were destroyed in an air raid in 1945, and we are thus unable to produce an accurate drawing. For the record, the essential data is given here:
Weight: 1,350,000 Kg (1329 tons) Length overall: 141 feet (42.976m)
Barrel length: 106 ' 6.7 " (32.480m) Elevation: 65° Traverse: Nil
Shell weight: 4.72 tons (4800 Kg) Velocity: 2690 f/sec (820 m/sec) or 7.0 tons (7100 Kg) Velocity: 2330 f/sec (710 m/sec) Maximum range: 51,400 yds (47,000m) or 41,560 yds (38,000m).

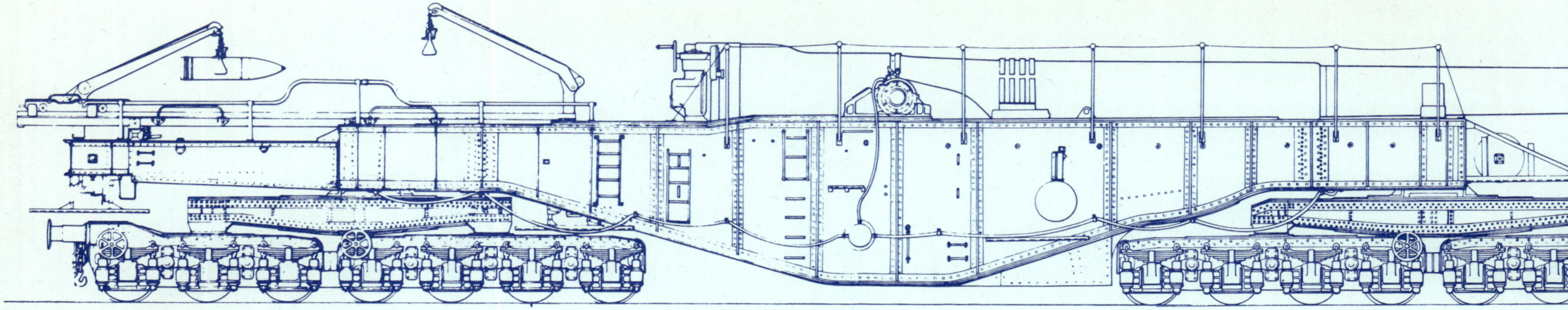

49 p.40

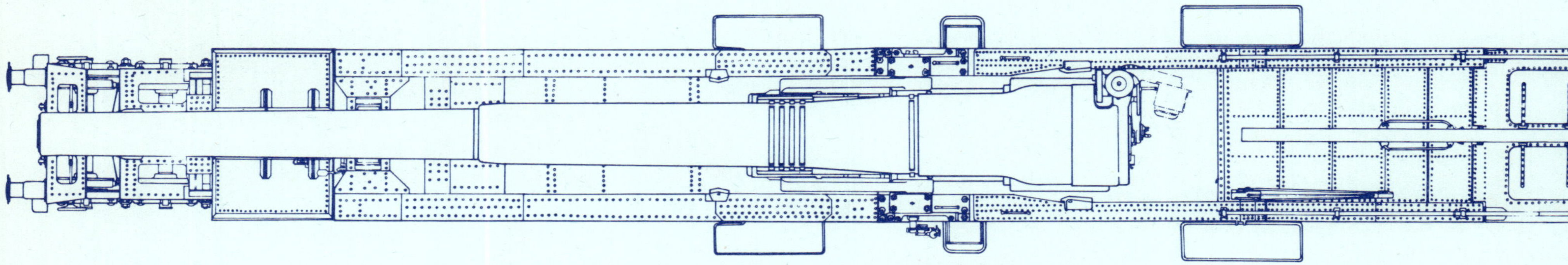

62 p.50

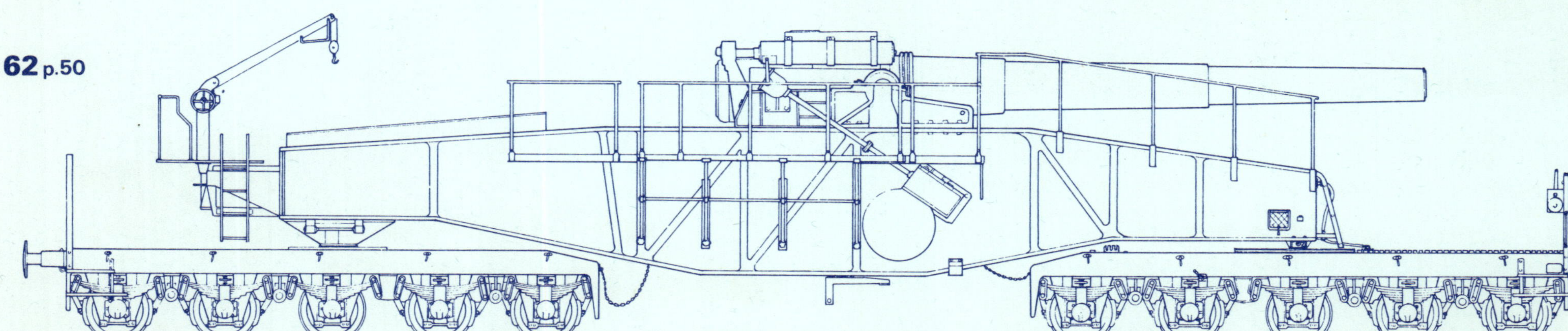

SCALE: HO

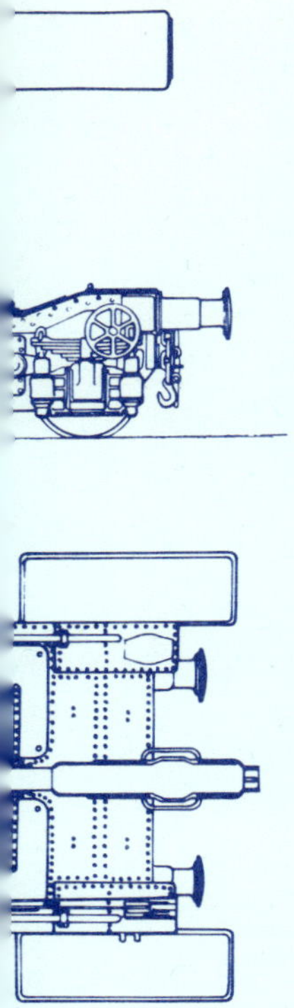

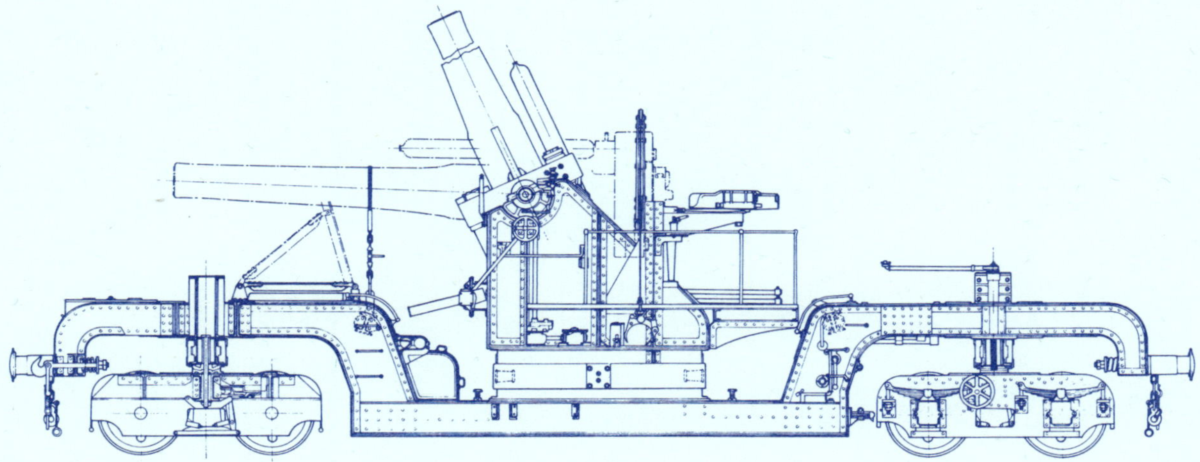

46 p.34

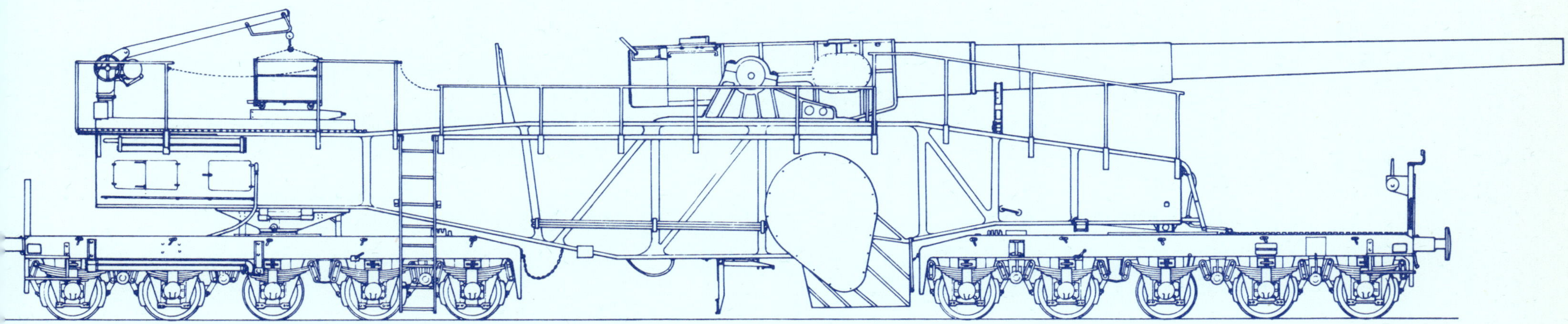

63 p.50

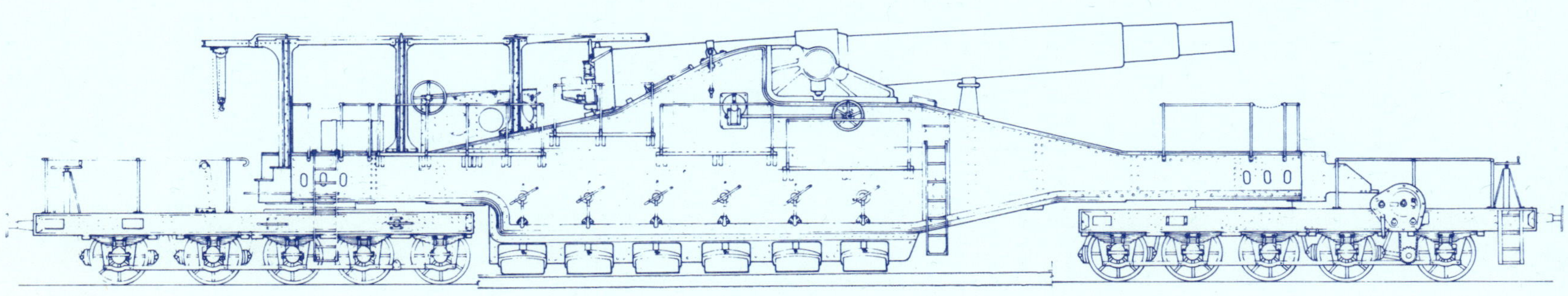

33 p.23

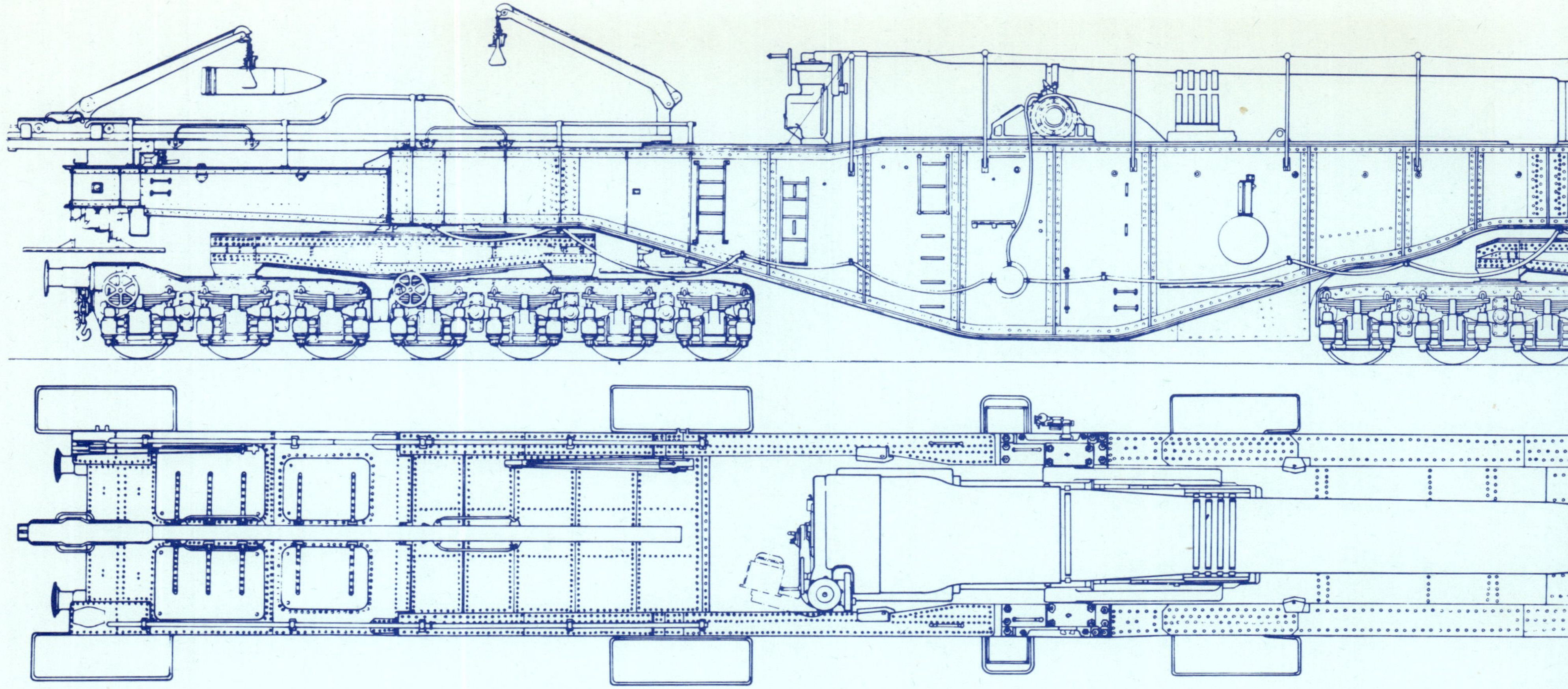

49 p.40

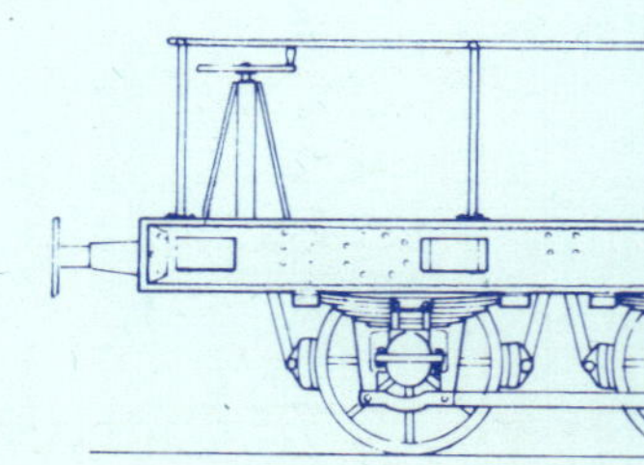

62 p.50

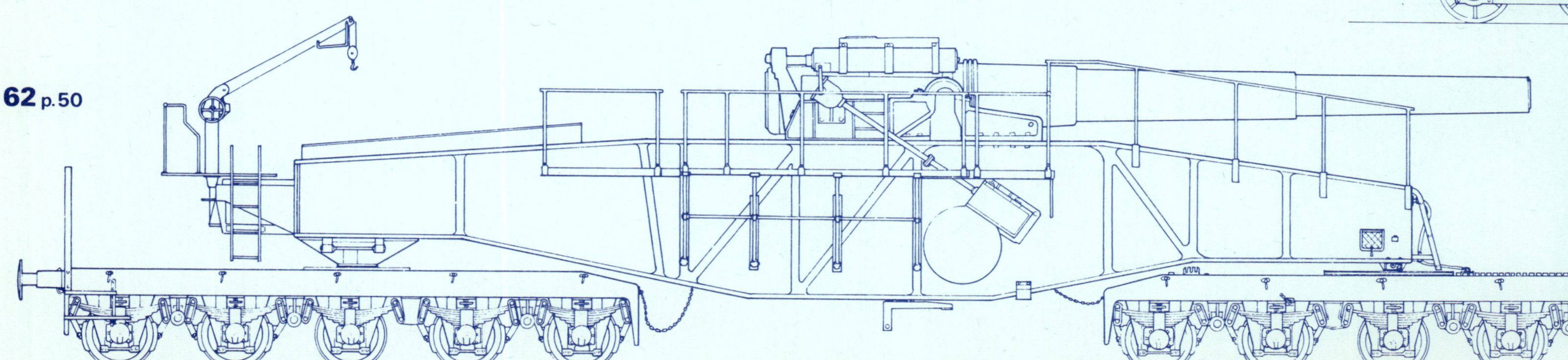

SCALE: OO

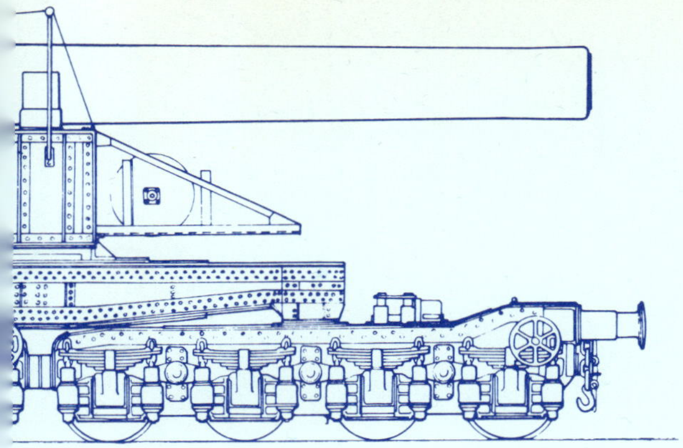

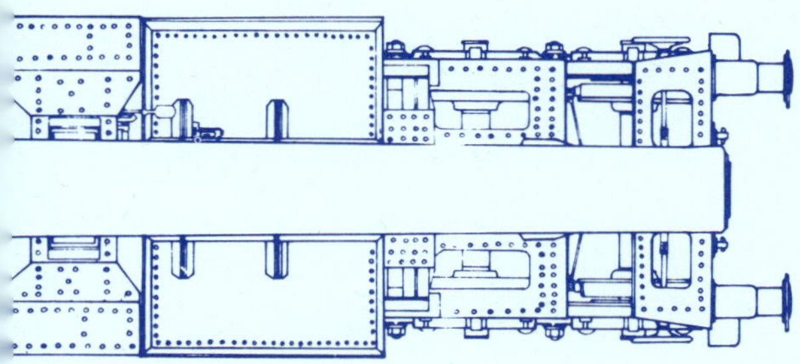

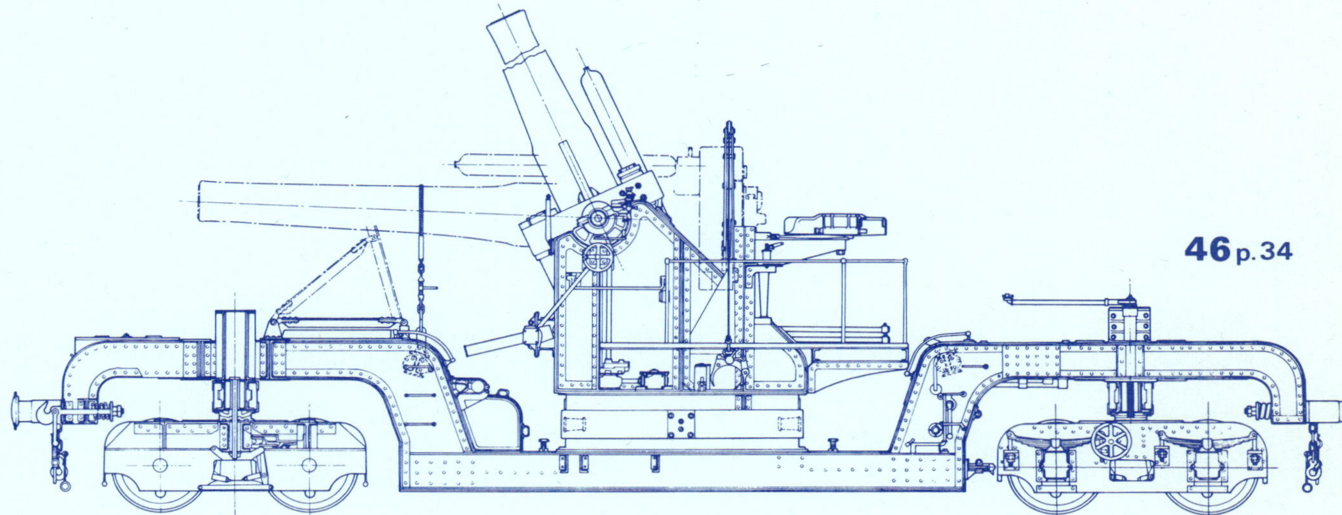

46 p.34

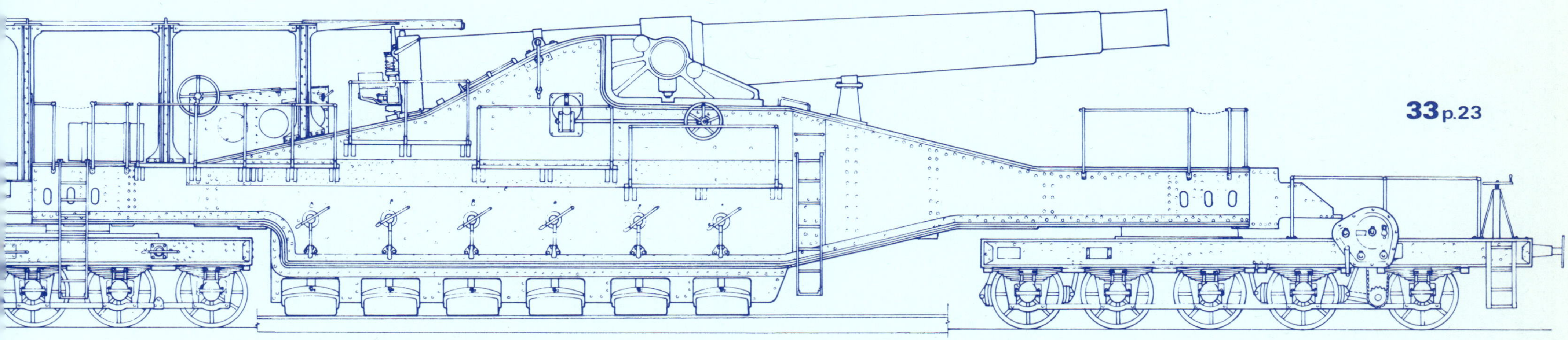

33 p.23

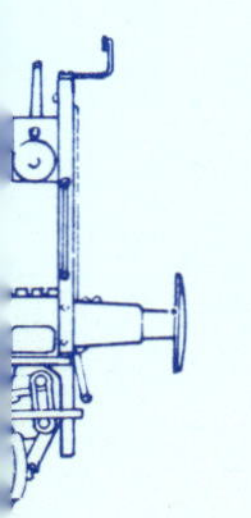